Business Process Improvement Through E-Collaboration:

Knowledge Sharing Through the Use of Virtual Groups

Ned Kock
Texas A&M International University, USA

IDEA GROUP PUBLISHING
Hershey • London • Melbourne • Singapore

Acquisitions Editor: Renée Davies
Development Editor: Kristin Roth
Senior Managing Editor: Amanda Appicello
Managing Editor: Jennifer Neidig
Copy Editor: Sue VanderHook
Typesetter: Sara Reed
Cover Design: Lisa Tosheff
Printed at: Integrated Book Technology

Published in the United States of America by
Idea Group Publishing (an imprint of Idea Group Inc.)
701 E. Chocolate Avenue, Suite 200
Hershey PA 17033
Tel: 717-533-8845
Fax: 717-533-8661
E-mail: cust@idea-group.com
Web site: http://www.idea-group.com

and in the United Kingdom by
Idea Group Publishing (an imprint of Idea Group Inc.)
3 Henrietta Street
Covent Garden
London WC2E 8LU
Tel: 44 20 7240 0856
Fax: 44 20 7379 3313
Web site: http://www.eurospan.co.uk

Library of Congress Cataloging-in-Publication Data

Kock, Ned F., 1964-
Business process improvement through e-collaboration : knowledge sharing through the use of virtual groups / Ned Kock.
p. cm.
Summary: "This book focuses on business process improvement, a key element of the most influential management movement since the 1980s, and how process improvement affects organizational knowledge sharing"--Provided by publisher.
Includes bibliographical references and index.
ISBN 1-59140-357-X (hardcover) -- ISBN 1-59140-358-8 (soft cover) -- ISBN 1-59140-359-6 (ebook)
1. Reengineering (Management) 2. Virtual work teams. 3. Knowledge management. 4. Organizational learning. I. Title.
HD58.87.K635 2005
658.4'022--dc22

2005004520

British Cataloguing in Publication Data
A Cataloguing in Publication record for this book is available from the British Library.

Business Process Improvement Through E-Collaboration:
Knowledge Sharing Through the Use of Virtual Groups

Table of Contents

Foreword vii

Preface ix

Chapter I
Introduction 1
Management Thinking and E-Collaboration 1
Business Process Improvement 3
Knowledge Explosion and Specialization 4
Business Process Improvement as a Catalyst for Knowledge Sharing 5
E-Collaboration Tools and Business Process Improvement Group Performance 6
The E-Collaboration Paradox 7
Increasing the Chances of Success of E-Collaboration Technology-Supported Groups 9
Summary and Concluding Remarks 10

Chapter II
Some History 11
An Historic View of Organizational Development 11
An Historic View of E-Collaboration 21
Summary and Concluding Remarks 29

Chapter III
What is a Business Process? 32
The Pervasiveness of Business Process Improvement 32
What is a Business Process?: Different Views 36
Summary and Concluding Remarks 49

Chapter IV
Data, Information and Knowledge 51
Data, Information and Knowledge are the Same Thing, Aren't They 51
Data are Carriers 55
Information is Descriptive 56
The Value of Information 58
Knowledge is Associative 60
The Value of Knowledge 63
Linking Data, Information and Knowledge 67
Summary and Concluding Remarks 70

Chapter V
Business Process Improvement and Knowledge Sharing 73
Organizational Knowledge and Competitiveness 73
The Need for Knowledge Sharing 74
Organizational Learning and Knowledge Transfer 77
Types of Exchanges in Organizational Processes 77
Business Process Improvement and Knowledge Communication 80
Summary and Concluding Remarks 86

Chapter VI
The Effects of E-Collaboration Technologies on Groups 89
Why Distributed Improvement and Learning? 89
Efficiency Effects from a Group Perspective 93
Effects on Group Outcome Quality 104
Effects on Knowledge Sharing Effectiveness 107
Further Evidence from Follow-Up Studies 111
Summary and Concluding Remarks 117

Chapter VII
The E-Collaboration Paradox 120
Paradoxical Results *120*
Research on E-Collaboration *121*
The E-Collaboration Paradox *123*
Media Naturalness: Human Beings Have Not Been Designed for E-Collaboration *124*
Compensatory Adaptation: Human Beings Often Try to Compensate for Obstacles Posed to Them *126*
Evidence from an Empirical Study of Business Process Redesign Pairs *127*
Summary and Concluding Remarks *130*

Chapter VIII
Successful Business Process Improvement through E-Collaboration 132
The Ubiquity of Business Process Improvement *132*
The Trend Toward Organizational Learning *136*
The Emergence of Virtual Organizations *138*
Success Factors: An Analysis of Twelve Groups *140*
Membership Factors *145*
Incremental or Radical Improvement? *154*
Can E-Collaboration Technology Support be a Trap? *155*
Summary and Concluding Remarks *158*

Chapter IX
Some Realistic Recommendations for Organizations 161
Information and Knowledge Explosion *161*
Distributed Improvement and Learning *164*
Some Recommendations for Organizations *167*
Popular Beliefs and Not So Popular Realities *169*
Organizational Culture Transformation through Education *171*
Summary and Concluding Remarks *172*

Chapter X
Using MetaProi to Improve Business Processes 175
MetaProi at a Glance *175*
Group Roles in MetaProi *178*

General Guidelines for MetaProi 180
Activities in MetaProi 181
MetaProi in Practice: A College Example 198
Summary and Concluding Remarks 211

Chapter XI
A Close Look at Twelve Business Process Improvement Groups **213**
A Structured Description of Several Groups 213
MAF.G1: Software Support 215
MAF.G2: The Internal Newsletter 217
MAF.G3: Pest and Disease Outbreaks 219
MAF.G4: Quality Management Consulting 221
MAF.G5: Information Technology Support 223
MAF.G6: Employee Training and Development 225
Waikato.G1: A Practical Computing Course 228
Waikato.G2: Academic Advice for Students 230
Waikato.G3: Student Computer Support 232
Waikato.G4: Student Assignments 235
Waikato.G5: International Graduate Students 237
Waikato.G6: New International Students 241

Appendix: Statistics for Those Who Hate Statistics **244**
The Importance of Statistical Tests 244
Comparing Means from Different Conditions 245
Checking for Correlations Between Variables 248
Assessing the Significance of Distribution Trends 251
Summary and Concluding Remarks 254

References **256**

Glossary **275**

About the Author **280**

Index **281**

Foreword

In today's organizations, a significant amount of work is conducted by virtual teams. While this provides a number of benefits to organizations and individuals, it creates a plethora of issues that are not always easily resolved. Virtual groups supported by collaborative technologies are blazing a path to new organizational forms. In this period, individuals have to develop new skills to deal with changes in work practices and manage organizational transformation.

Ned Kock's book explains how organizations can be redesigned, facilitated by collaborative technologies. His work over recent years has led to him coining the term "e-collaboration" to describe the rapidly emerging area where information and communication technologies support distributed teams. Although the technology is developing in sophistication, on a yearly basis the organizational and management practices to manage the new ways of working are stumbling. An example of this awkwardness in effectively employing collaborative systems is highlighted with the use of e-mail in organizations. Many employees are suffering from information overload, and e-mail is a prime contributor.

Kock explains the history of business process improvement and its importance in the current business landscape. His aim is not to just look at the implications of collaborative technologies, but also to examine how they can be best employed for organizational productivity improvement.

Knowledge management is a huge topic with many facets, but most people concerned with organizational productivity would agree that it is important.

Kock's work explains where and how knowledge can be effectively shared and managed in the business process improvement cycle. Of course, to do this requires the skills and knowledge to effectively employ collaborative technologies in organizations. It is particularly in this area that Ned excels.

I encourage you to read the book, reflect on its contents, and put into practice initiatives based on the insights from the in-depth research that Ned has conducted over a number of years. Finally, I would like to congratulate Ned on an excellent contribution to the literature on this topic.

Dr. Craig Standing
Editor-in-Chief, Journal of Systems and Information Technology
Professor of Strategic Information Management
Edith Cowan University, Australia

Preface

Purpose of This Book

This book is based on my previous book titled *Process Improvement and Organizational Learning: The Role of Collaboration Technologies* (1999, Idea Group Publishing). Among the revisions (some extensive) that led to this new book is the inclusion of evidence from studies of business process improvement and e-collaboration issues conducted in the US in the past seven years. In most cases, the new evidence has been added to and integrated with the evidence already summarized in the previous book. Here, I discuss several new issues regarding the impact that e-collaboration technologies have on business process improvement and knowledge sharing.

I have written this book around two main theses. The first is that business process improvement, a key element of the most influential management movements since the 1980s, can be considerably improved by the use of information technology. I argue that distributed and asynchronous e-collaboration technologies; such as e-mail, computer conferencing, and Web-based groupware systems, are likely to play a major role in this improvement.

The second thesis set forth in this book is that business process improvement affects organizational knowledge sharing (a key component of what is generally referred to as organizational learning) in a non-linear way, and that the use of e-collaboration technologies can boost this influence by increasing the breadth and speed of knowledge dissemination in organizations. To lay the groundwork for the development of this thesis, I explore the relationship among the concepts of data, information, and knowledge. I do so by looking at how these abstract entities affect our lives. These concepts are then used as the building blocks to define knowledge sharing and organizational learning.

This book is primarily based on my experience facilitating and researching over 100 business process-focused organizational development projects, including a three-year project where more than 38 process improvement groups were facilitated with the support of e-collaboration technologies in three different countries. Here, I look into the relationship between e-collaboration technology use, business process improvement success, and knowledge sharing effectiveness.

Key findings in connection with effects of e-collaboration technologies on business process improvement groups are discussed in this book. Among those findings are the following: the use of e-collaboration technologies appears to increase the number of simultaneous business process improvement groups that can be conducted in an organization; it also seems to decrease the organizational cost of business process improvement groups; and, moreover, it appears to have a neutral (or, at least, non-negative) effect on the quality of the outcomes generated by business process improvement groups.

A related finding regarding knowledge sharing is that the use of e-collaboration technologies seems to have a positive impact on how much knowledge is disseminated across the organization, and how fast that happens when those technologies are used in combination with catalyst efforts such as business process improvement.

The above findings may seem a bit academic at first glance, but they have vast applications in business. For example, improvements regarding business process improvement efficiency mentioned above can be translated into significant savings in the implementation of business efforts aimed at quality and productivity improvement, such as the following:

- Quality management certification programs based on standards by the International Organization for Standardization (ISO) like the ISO 9000 family of standards.
- Industry-specific accreditation efforts, such as those that universities and colleges undergo to obtain accreditation of their educational programs by the Association for Advance Collegiate Schools of Business (AACSB) and the Southern Association of Colleges and Schools (SACS).
- Process-specific certification efforts, such as those undergone by large information technology defense contractors and other software develop-

ment organizations that bid for large contracts, in connection with the Software Engineering Institute's Capability Maturity Model (CMM), which classifies information technology organizations at levels 1 to 5 (5 being the best) according to their adherence to CMM's prescriptions.

- Quality improvement efforts based on specific methodologies such as the Six Sigma methodology, to which much praise has been directed recently by the leaders of large and influential companies such as Honeywell, General Electric, and Lockheed Martin.

The findings summarized above suggest that the use of e-collaboration tools, if done properly, can reduce the cost and increase the speed of efforts aimed at ISO 9000 and CMM certification, AACSB and SACS accreditation, and Six Sigma implementation. And these are just a few of the certification and accreditation schemes that have found widespread adoption in particular industries.

Moreover, the specific finding that the use of e-collaboration technologies seems to have a positive impact on the speed and breadth of organizational knowledge dissemination may be put into practice to support key business efforts that are becoming increasingly common, such as knowledge transfer between subsidiaries of the same company, whenever new technologies and/or methods are developed; parent and acquired businesses in post-merger situations; and main company and contractor, in strategic outsourcing partnerships.

However positive the above findings may sound, they can only become reality if some precautions are taken by organizations. This book discusses several of these precautions and lays out a blueprint to conduct e-collaboration technology-supported business process improvement and knowledge sharing. The book also includes a detailed description of a tested methodology to guide the work of e-collaboration technology-supported business process improvement groups.

From a broad perspective, my goal with this book is to help managers, as well as students who are pursuing a management career, to prepare their (future and present) organizations to survive and thrive in the Internet era. This is the era where, more than ever, the fittest organizations are those able to master the art of efficient and effective acquisition and use of data, information, and knowledge.

Organization of This Book

This book is made up of 11 chapters, one appendix, a reference section, and a glossary. The content presented in the chapters flows in such a way to introduce the reader to fundamental ideas, to develop and support with evidence the two basic theses of this book, and finally, to offer some advice to organizations on how to implement them. Chapters I through IV are more introductory in nature. The remaining chapters build on extensive empirical evidence, mostly collected in previous projects led by the author. The appendix provides a basic reference on statistical analysis techniques for those who "hate" statistics. Below is a summary of each chapter's content.

Chapter I offers an introduction and motivation for the book. It introduces several topics that will be discussed in more detail in later chapters. Among those topics are the evolution of management ideas and the related evolution toward e-collaboration technologies, the roots of business process improvement ideas, the trend toward knowledge fragmentation, the role of business process improvement with regard to knowledge sharing, the effect of e-collaboration technology support on business process improvement groups, the e-collaboration paradox, and the success factors associated with e-collaboration technology-supported business process improvement groups.

Chapter II offers a historical review of the fields of organizational development and e-collaboration. This review focuses on major historical events and does not restrict itself to academic issues. As such, several of the major management developments are discussed, from Adam Smith's division of labor approach in the 1700s to Hammer and Champy's reengineering movement in the 1990s. Subsequently, the chapter describes the main technological developments that led to the emergence of e-collaboration tools as significant tools for organizational improvement and organizational learning. In this chapter, I attempt to build a link between the commercial establishment of computing technologies and the organizational development ideas that became popular in the same period. This should help readers understand how technologies that enable e-collaboration have evolved vis-à-vis the development of major management ideas, and get situated in the topical discussion that will be expanded in further chapters.

In spite of the recent popularity that the business process concept has gained, I believe it is a fundamental idea behind several previous management movements, including the total quality management, organizational learning, and business process reengineering movements. Therefore, I use Chapter III to discuss the concept of business process in the context of the management movements just mentioned. I also describe several popular views of processes, with particular attention to the data and workflow views.

Chapter IV is also an introductory chapter. It discusses three fundamental concepts referred to throughout the book—data, information, and knowledge. This chapter is particularly important because of the rather confusing way in which these terms are used in the academic as well as the more popular senses. Here, I offer new conceptualizations that suggest that data is a carrier of information and knowledge, and that while information is eminently descriptive, knowledge's nature is mostly predictive. Although these conceptualizations are heavily based on previous theoretical frameworks from cognitive science and artificial intelligence, I try to eliminate technical jargon as much as possible and explain my views through examples involving simple day-to-day situations.

Chapter V presents and discusses empirical evidence that supports one of the core theses of this book, which is that business process improvement, when viewed as a meta-process, affects knowledge sharing in a positive way (the term "meta" is used to refer to a high-level process through which process improvement is sought). This chapter does not discuss any direct effect that e-collaboration technologies may have on people (this is done later). Rather, it targets a specific group process—the business process improvement meta-process—and shows that this group process (with or without technology support) leads to increased knowledge communication and sharing in organizations.

An extensive discussion of the direct and indirect effects that e-collaboration technologies are likely to have on business process improvement and knowledge sharing is provided in Chapter VI. This is one of the core chapters of the book and addresses both of the two main theses of this book. Most of the chapter is about e-collaboration technology effects in connection with business process improvement. The emphasis is on asynchronous e-collaboration technologies, particularly those based on the electronic mail paradigm. At the end of the chapter, a discussion about the impact on knowledge sharing is also presented. This discussion builds on evidence presented earlier in the book, as well as new evidence introduced in this chapter.

Chapter VII provides a discussion of the results summarized in the previous chapter, and puts forth a curious picture that seems to plausibly explain those results (which would appear to be somewhat puzzling at first glance). The picture is characterized by two competing phenomena associated with the use of e-collaboration technologies; phenomena that are collectively referred to here as the "electronic collaboration paradox."

(a) People seem to consistently perceive face-to-face communication (as well as communication that incorporates key elements of the face-to-face medium, such as the ability to use non-verbal cues like tone of voice and body language to convey ideas) to pose fewer obstacles to effective communication than other, particularly electronic, media.

(b) When groups conduct collaborative tasks using e-collaboration technologies, they often present the same level of performance or even perform better than groups accomplishing the same tasks face-to-face, which is contradictory with notion (a).

The curious phenomena above are presented as reasons why e-collaboration technologies should be used according to certain prescriptions (discussed later in the book), since their unwise utilization may lead to more problems than benefits.

While the evidence presented in the previous chapters suggests an overall positive impact of e-collaboration technologies on business process improvement and knowledge sharing, it does not address success factors related to the use of e-collaboration tools to support business process improvement and knowledge sharing groups. This is done in Chapter VIII, based on the discussion previously presented in connection with behavioral effects of e-collaboration technologies. This chapter presents a careful success/failure analysis of several e-collaboration technology-supported groups, which leads to the identification of a few critical success factors. The chapter discusses the appropriateness of e-collaboration technology support in incremental and radical business process improvement situations, and shows that e-collaboration technology support may become a trap to organizations if not properly employed.

Chapter IX summarizes the findings discussed in previous chapters, particularly those presented in connection with behavioral effects of e-collaboration technologies, and whose discussion is expanded in later chapters. This summarization relies in part on a graphical model depicting an integrated view of e-collaboration technology support effects on distributed business process improvement groups and knowledge sharing. This chapter also presents a number of "realistic" recommendations for organizations trying to avoid much of the self-servicing advice often seen in popular business publications.

Chapter X provides a detailed description of MetaProi, the group methodology I used to facilitate several of the groups analyzed in this book. I expect this chapter to be instrumental in similar future initiatives by the readers. MetaProi is a new approach for business process improvement, which by necessity is based on previous books and papers on the topic. MetaProi's focus is both on the quality *and* productivity of business processes. The methodology was designed so that it can be conducted through electronic as well as face-to-face meetings. One detailed example of an electronic discussion based on MetaProi is provided at the end of this chapter.

Chapter XI provides a summarized and structured description of 12 cases of e-collaboration technology-supported business process improvement, where e-collaboration is utilized at different degrees (in some cases, together with face-to-face collaboration). The e-collaboration tool used is an electronic mail system with list-distribution capabilities, which is ideal for the purposes of this book because it represents a widely used and relatively cheap type of e-collaboration technology that just about any organization can easily adopt. Since these cases served, together with other cases and experiences, as the basis for several of the ideas in the book, I think that they will also be useful to readers who want to draw additional conclusions based on raw data.

The appendix discusses several statistical analysis techniques used in the book. The goal of this appendix is to explain, in simple terms, what the several statistical analysis techniques mentioned in the book mean. While mentioning the techniques in other chapters of the book lends credibility to the arguments presented there, and provides a basis on which experts can evaluate those arguments, it can also lead to some confusion, especially among readers who are not very familiar with those statistical analysis techniques. This appendix is an "antidote" against that potential confusion. Some of the statistical tests that are briefly covered include comparison of means tests (e.g., the T test), correlation tests (e.g., the Pearson correlation test), and distribution trend significance tests (e.g., the Chi-squared goodness-of-fit test).

Acknowledgments

Several sections of Chapter 7, where the e-collaboration paradox is discussed, have been adapted from a previously published article titled "Resolving the E-Collaboration Paradox: The Competing Influences of Media Naturalness and Compensatory Adaptation." The article was co-authored by me, and appeared in the German journal *Information Management and Consulting* (Special Issue on Electronic Collaboration, V.17, No.4). Its sections have been adapted and used here with permission by the journal's publisher.

I would like to acknowledge the invaluable cooperation of the many organizations that have been involved in the projects that led to this book. Those organizations include HSBC Bamerindus Bank, PricewaterhouseCoopers, Johnson & Johnson, Rio de Janeiro State Construction Company, Westaflex, New Zealand Ministry of Agriculture and Fisheries, True North, Accenture, Metro One Communications, Marriott Hotels, Day and Zimmermann, Lockheed Martin, and Bristol-Myers Squibb.

Part of the evidence discussed in this book was collected and analyzed through funded research projects. I am particularly indebted to the US Defense Acquisition University for funding my research through their External Defense Acquisition Research Program. Funding and time release was also provided by Texas A&M International University's College of Business Administration; special thanks go to Dr. Steve Lunce, who served as my department chair at the time of writing.

This book has its roots in my Ph.D. research project, conducted several years ago at the University of Waikato, in New Zealand. My special thanks go to my former Ph.D. dissertation advisor, Bob McQueen, and his family. Bob has not only provided excellent academic support throughout my Ph.D. research, but also a sincere friendship and advice about general life matters. He and Anita McQueen have gone out of their way to make my family and me feel as comfortable and at home as possible while I was working my way toward the completion of my Ph.D. degree program.

Thanks are also due to the many close friends, former students (some of whom are also close friends), and colleagues who challenged several of my assumptions in connection with the main topics of this book and who, by doing that, helped me shape several of the ideas that I proposed and discussed here. I would particularly like to acknowledge the following individuals: Robert Aiken, Richard Baskerville, John Buchanan, Kim Campbell, Azim Danesh, Dorrie DeLuca, Darleen DeRosa, Kevin Dow, Roberto Evaristo, Frank Friedman, Donald Hantula, Pedro Hurtado, Giorgio Ingargiola, Allen Lee, Gavan Lintern, Jim Maskulka, Michele Masucci, Fred Murphy, Michael Myers, John Nosek, Ronaldo Parente, Elisabeth Rossen, Craig Standing, Jacques Verville, and Ron Weber.

Last but not least, I would like to thank my family. I dedicate this book to them.

Ned Kock

Chapter I

Introduction

Management Thinking and E-Collaboration

One cannot help but admire the accomplishments achieved by organizations over the years, and their impact on society in general. While some may argue that we have become too urbanized and materialistic, it is undeniable that today most of us can buy many more goods and services than our great grandparents could (at a lower price relative to inflation), which seems to make our life easier and more comfortable. Of course, that makes us more prone to buy things that we don't need, and feel unhappy about the results. But that is not the fault of progress; we should blame our inability to deal with progress.

Underlying the above accomplishments are management ideas, which have over time been associated with management schools of thought that received specific names (e.g., scientific management and total quality management). When one looks at how management schools of thought evolved, from the early specialization and division of labor ideas in the late 1700s to the emergence of the humanist movements in the early and mid-1900s, that evolution seems to have a particular direction. Work has been structured more and more around group collaboration.

Interestingly, the idea that successful group collaboration is important for organizational performance was quite well established already in the mid-

1900s during the post World War II period. Technologies were available to support e-collaboration, mostly in the form of mainframe-based systems. However, the use of e-collaboration tools to support group work in business took some time to catch up, often because the cost of computer technologies was too high for those technologies to be used by anyone other than nerdy types (often called the "computer folks") working in central data processing departments.

The above situation, of course, has changed substantially over time, especially after the 1980s, when there was an explosion in the use of computer networks, and when computer equipment became relatively cheap, allowing for increasingly decentralized access to computer resources. The nerdy types lost their monopoly on computer resources, and many central data processing departments were dissolved and replaced by information technology support departments, most of which today are big help desk operations.

Interestingly, the above mentioned loss in monopoly was accompanied by a slowly increasing involvement of computer experts in highly successful entrepreneurial endeavors, which led many of the previously called nerdy types to become extremely wealthy and assume different organizational titles, such as president and chief executive officer (CEO)—does the name Bill Gates ring a bell?

With the explosive growth of the Internet and the Web in the 1990s, most computers became interconnected, which led some to see the computer as less of an autonomous processing unit and more of an entry point to a vast pool of network-based resources. The increasing use of e-collaboration technologies led to many possibilities; one of the most exciting was the ability to conduct collaborative tasks interacting at different times (i.e., in an asynchronous manner) and from different places (e.g., different cities or countries).

This book focuses on e-collaboration technologies that enable group-based interaction at different times and from different places, and the impact that those e-collaboration technologies have on business process improvement groups. The term *e-collaboration* is used here as an umbrella term that comprises several other closely related fields, commonly known as computer-mediated communication, computer-supported cooperative work, groupware, group support systems, collaboration technologies, or, more recently, the so-called field of knowledge management.

Business Process Improvement

The idea of looking at how things are done in organizations (with the goal of changing them as a basis for improving organizational performance) has been around for a long time. Taylor's own scientific management approach, which he developed and implemented in the late 1800s and early 1900s, arguably can be categorized as focusing on the how-to of production; that is, on how things are done. This how-to perspective on organizational development later got a name—the business process-oriented perspective.

William Edwards Deming, the father of the total quality management, hammered the nail on the head when he emphasized the notion that quality improvement should be based on a focus on business processes, not people or products, because most product-related quality problems are caused by poor business process design. For example, if the steps that a worker has to carry out to produce a computer card are poorly designed, then problems will occur, such as a high proportion of defective computer cards. There is no point in trying to focus on what's wrong with the workers in situations like this. While there certainly are exceptions, most of the quality problems are not the workers' fault. Deming later became famous for applying Pareto's rule to quality improvement, by arguing that 80% of product quality problems are business process-related, whereas only 20% can be blamed on the people who carry out business process activities and other factors (Deming, 1986).

While Deming hammered the nail on the head, it took a Hammer—that is, Michael Hammer—to really make everyone in business pay close attention to business processes and their redesign as a way of improving organizational performance. Not only did Hammer address the issue of how to improve organizations, but he also dealt with the nagging issue of how to get a decent return on investment in computer technologies.

Although investment in computer technologies has certainly improved manufacturing productivity, it often has led to disappointing results in the service sector, particularly before the 1990s. Michael Hammer, along with others (notably Champy, Davenport, and Short), suggested in the early 1990s that business process redesign was the key to obtaining substantial returns on information technology investment.

The jury is still out on whether Hammer was right, especially because the ideas that he proposed, grouped under the business process reengineering banner, were later found to be difficult to implement successfully (Champy, 1995;

Kock, 2003). Many have found business process reengineering difficult to implement because of its disruptive nature, which calls for radical organizational changes. This has led to much criticism and a high rate of failure for reengineering projects.

Notwithstanding the fact that business process reengineering was broadly criticized for a high failure rate, it is reasonable to assume that all the discussions surrounding reengineering have led to a renewed focus on business processes and their redesign, and the integration of that focus with other foci coming from less radical organizational improvement schools of thought, such as total quality management (Davenport, 1993a). As a result, business process improvement has become much more pervasive. And, as often happens, consulting companies cashed in by developing services around the notion of business process improvement.

Was this trend bad for the economy? If that had been the case, one would expect to see an overall decrease in organizational performance, which is not what happened. Not only did organizational productivity continue to increase, but there also has been evidence that returns on investment in computer technology have gone up since the mid-1990s.

Is there a causal relationship here of this type: business process improvement pervasiveness → increases in organizational productivity improvement? I personally think so, and this is one of the beliefs that made me write this book.

Knowledge Explosion and Specialization

One other thing that seems to be happening lately, and that has been felt more acutely since the 1990s, is the explosion in the amount of knowledge and information flowing in organizations of all kinds. This is particularly true in the service sector, which is the largest contributor to any developed and developing country's gross national product.

Organizational knowledge is predominantly stored in people's brains. That is, organizational knowledge is largely an aggregate of the knowledge held by the individuals who work in business processes, even though knowledge can also be stored in documents, databases, and software.

The problem here is that the more knowledge that is generated, the more specialization occurs. That is, the increasing volume of existing knowledge in all areas pushes knowledge holders into specialization. This leads to knowledge fragmentation in organizations, which, in turn, has been attacked by organizational learning advocates as a key reason for low business process productivity and quality. The solution proposed by some management gurus simply is to stimulate interfunctional knowledge sharing.

That is easier said than done. One of the alternatives to stimulating interfunctional knowledge sharing proposed in this book is to conduct business process improvement in a continuous and organization-wide manner as a basis of organizational improvement programs.

Business Process Improvement as a Catalyst for Knowledge Sharing

One can stimulate interfunctional and cross-departmental knowledge sharing in different ways. For example, organizational governance committees and training retreats can be used to bring together people from different organizational areas to discuss issues that will lead to knowledge sharing. This is a traditional approach to stimulate knowledge sharing.

The problem with many of the traditional approaches to stimulate knowledge sharing, in my opinion, is that their goals are often a little too ethereal for people to get excited about. Many employees that I know in a variety of organizations think very little of retreats and governance committees (unless they are very high-level committees, such as a board of directors or trustees). Those retreats and committees are often seen as a big waste of time.

Couldn't knowledge sharing be a by-product of a business process that had some direct appeal to managers and employees, as well as some impact on the organization's bottom-line? If the answer is yes, then what would be such a process? The process (or meta-process) of business process improvement comes to mind. While this idea has been hinted at in the past, the first studies that looked at business process improvement's role as a catalyst for knowledge sharing were conducted by Kock (1997) (see also Kock & McQueen, 1998a; 1998b; Kock & Davison, 2003).

These studies discuss evidence (also discussed later in this book) that suggests that the number of knowledge-bearing communication exchanges in business process improvement meta-processes is much higher than that observed in routine (or non-improvement) processes. The business process involved in shipping an order of books to a customer is an example of what I call a routine (or non-improvement) process. Conversely, the business process involved in improving the process of shipping an order of books to a customer is an example of what I call a business process improvement meta-process.

The evidence I obtained from previous studies also indicates that the proportion of knowledge content in communication exchanges in business process improvement meta-processes is approximately 35%, compared to approximately 15% for routine business processes.

These findings are quite interesting and serve as a basis on which business process improvement can be recommended for reasons that go beyond its main expected outcome (i.e., better business processes). In other words, business process improvement, unlike traditional knowledge transfer activities such as training sessions and committee meetings, has side effects that are obviously beneficial; among these are chiefly the improvement of business process productivity and quality.

E-Collaboration Tools and Business Process Improvement Group Performance

Let's add a new variable to the picture painted above, which alone indicates that business process improvement is a good thing, not only because of its inherent advantages, but also because it stimulates knowledge sharing. The new variable is e-collaboration technology support, which takes us full circle to the beginning of this chapter, when I discussed the emergence of the Internet and the Web, along with the related trend toward e-collaboration.

What does e-collaboration technology support do for business process improvement groups? While this question seems rather important to me, as a researcher, my review of the literature suggests that very few researchers tried to answer it based on systematic research. Among those researchers are Alan Dennis et al. (1999), as well as some of my colleagues and I (Kock, 2001; Kock & Corner, 1997; Kock & McQueen, 1998). Dennis et al. focused their analysis on e-collaboration systems normally called group decision support systems, which usually are employed to support groups meeting in the same room and at the same time. Those systems were found to reduce meeting time, especially in meetings involving brainstorming and decision-making tasks. My own research with colleagues focused on asynchronous e-collaboration systems, which support group work where members interact at different times and from different places.

Our research points to an increase in business process improvement group efficiency due to e-collaboration technology support. Those efficiency gains are primarily associated with reduced group cost, lifetime, and reliance on managers. Additionally, the evidence collected and compiled through our research suggests that knowledge sharing effectiveness, and the number of simultaneous business process improvement groups that can be conducted in an organization, are both increased by e-collaboration technology support. As for group outcome quality, the evidence from our research points to a slight increase as a result of e-collaboration technology support.

In spite of the above findings, which are no doubt positive, our research also yielded evidence that the electronic communication medium generated by e-collaboration tools poses obstacles for communication in groups in general, and in business process improvement groups in particular. This is obviously difficult to reconcile with the generally positive outlook discussed above, and calls for a deeper theoretical explanation, which led me to develop a theoretical notion called the e-collaboration paradox (Kock & D'Arcy, 2002).

The E-Collaboration Paradox

The e-collaboration paradox is characterized by two general and competing findings in connection with the impact of e-collaboration tools on groups. The first finding is that group members generally perceive face-to-face communication as posing fewer obstacles to effective communication than other communication media, particularly media generated by e-collaboration systems. The reason seems to be that e-collaboration systems remove important elements present in the face-to-face medium, such as the ability to use non-verbal cues to convey ideas.

The second finding is that when groups conduct collaborative tasks using e-collaboration systems to support interaction among group members, those groups often present the same or higher levels of performance as groups in which members interact primarily face-to-face. This second finding clearly contradicts the first finding.

I will posit in more detail later in this book that the e-collaboration paradox can be easily explained based on two hypotheses. The first hypothesis puts forth the notion that the biological communication apparatus of human beings has been designed by Darwinian evolution to excel in face-to-face communication, and that consequently, e-collaboration tools that eliminate too many of the elements present in face-to-face interaction end up making communication more difficult, or cognitively demanding. The reason for this is that human beings spent most of their evolutionary history, which spans millions of years, communicating primarily face-to-face. Writing is a very recent phenomenon, having emerged about 3,000 B.C. in the Sumerian culture, and even cave painting is a recent phenomenon, dating back to about 40,000 B.C. I like to think of this hypothesis as the "ape that used e-mail" hypothesis (Kock, 2001c).

The second hypothesis puts forth the notion that that we humans invariably adapt our communicative behavior when faced with communication obstacles, including obstacles posed by e-collaboration technologies, which often leads to an interesting and somewhat puzzling result—we perform just as well or even better than if those obstacles were not present. This happens even though most of us still perceive e-collaboration tools as generally posing cognitive obstacles to communication. That is, the obstacles end up leading to success, albeit in an indirect way.

I was so interested in this compensatory adaptation phenomenon that not only did I study it in the context of e-collaboration tools (Kock, 1998, 2001b), but I also decided to look into it from a broader perspective. As a result of that decision, I wrote a book (published in 2002) based on case studies that investigate compensatory adaptation as a general human phenomenon. That book provides examples in several diverse areas of business and society, suggesting that compensatory adaptation is, indeed, a very general human phenomenon that can explain adaptations that go well beyond the scope of e-collaboration technology-induced adaptations in communication behavior (Kock, 2002c).

Increasing the Chances of Success of E-Collaboration Technology-Supported Groups

It seems, then, that e-collaboration technology support has some beneficial effects on business process improvement groups, and that those beneficial effects result from a complex phenomenon that involves some form of compensatory adaptation, since the communication media created by most e-collaboration technologies are not as natural to human beings as the face-to-face medium. Even advanced virtual reality technologies do not replicate all the elements present in face-to-face communication.

Yet, e-collaboration technology-supported business process improvement groups also fail. According to a recent study based on evidence collected in New Zealand and the US regarding the success rate of business process improvement groups interacting primarily electronically, about 62% of those groups succeeded. This is an encouraging result, since the literature on traditional (non-e-collaboration technology-supported) business process improvement reports success rates of improvement attempts based on total quality management principles ranging from approximately 20% to 34%. For improvement attempts employing reengineering principles, the success rate obtained from the relevant literature was approximately 30%.

In spite of this positive outlook, a 38% failure rate is enough to warrant the search for success factors in e-collaboration technology-supported business process improvement. The same research that provided us with the 38% failure rate figure also suggests three main types of success factors—leadership, membership, and other identified factors that are unrelated to leadership and membership. Leadership factors relate to characteristics of the leaders (or moderators) of e-collaboration technology-supported business process improvement groups. Membership factors relate to group membership configurations. As for the other factors, those relate to general characteristics of each group, including characteristics of the target business process.

The analysis of success factors, conducted through the research mentioned above and discussed in more detail later in this book, allows us to describe an e-collaboration technology-supported business process improvement group whose likelihood of success would be very high. Such an "ideal" group would have the following characteristics: (1) its leader would be the "owner" of the target business process (i.e., the manager or main person responsible for the business process execution); (2) the group's discussion would present little personal risk for its members, and most of its members would have a personal stake in its outcome; and finally (3) the ideal group would target a relatively narrow and simple business process.

Summary and Concluding Remarks

The topics touched on here will be discussed in more detail later in this book. Among those topics are the evolution of management ideas and the related evolution toward modern e-collaboration technologies, the roots of the business process improvement idea, the trend toward knowledge fragmentation, the role of business process improvement in regards to knowledge sharing, the effect of e-collaboration technology support on business process improvement groups, the e-collaboration paradox, and the success factors associated with e-collaboration technology-supported business process improvement groups.

The discussion of some of these topics takes up entire chapters of this book. Other topics are discussed in combination within a single chapter. I hope that those readers who read through this entire book (even if they skip a few sections along the way) will have a clear picture of how business process improvement, a key element of the most influential management movements since the 1980s, can be considerably improved by the use of information technology. I also hope that the readers of this book will have a clear picture of how business process improvement can affect organizational knowledge sharing (a key component of a framework for organizational improvement, often referred to as organizational learning) in a non-linear way. Finally, I hope that e readers will understand how e-collaboration technologies can be used to boost the beneficial effects of business process improvement on organizational performance.

Chapter II

Some History

An Historic View of Organizational Development

Organizational development is the generic field of research and practice concerned with structural organizational changes that can have a positive impact on competitiveness. It is about changing organizations in order to make them more competitive, chiefly through modifications in their organizational structure. Historically, most organizational development efforts have aimed at improving productivity (i.e., cycle time and cost reduction) and quality (i.e., boosting customer satisfaction). Organizational development encompasses procedural and policy changes within firms in order to adapt to external factors. External factors include competitive pressures, as well as economic and government regulation changes.

The history of organizational development is closely linked with the history of management. Those who have initiated and championed organizational development approaches and ideas typically have also had the responsibility of coordinating the efficient and effective deployment of human and material resources in organizations as managers, management consultants, or action-oriented management researchers.

Although the development of management as an academic discipline is a relatively recent one, management procedures that resemble those commonly used today in large corporations can be traced back to as early as 5,000 B.C. At that time, the Sumerians, who invented the very first form of written language, also developed careful record-keeping procedures to keep track of cattle, stones used in construction, and other items. The Egyptians followed, with the development of planning and coordination procedures that are reflected in the precision with which they have accomplished a number of very complex construction projects, such as the large irrigation networks built around the Nile River and the Great Pyramids.

The expansion of the Roman Empire and the need to effectively manage occupied territories and provinces have seen the development of common measurement systems and standards to facilitate communication. This period also saw the development of job descriptions to ensure that people in management and administrative roles clearly understood what was expected from them.

In the Middle Ages, Italian merchants developed elaborate bookkeeping procedures, which included the introduction of the double-entry (debit-credit) bookkeeping approach by the Franciscan priest Luca Pacioli, the basics of costs accounting, and the concepts of journal entries and ledgers. In that same period, the benefits of task specialization, which preceded the concept of division of labor, and rudiments of strategic management were proposed by Thomas More in England and Niccolo Machiavelli in Italy, respectively.

Adam Smith, a professor at Glasgow University in Scotland, later picked up the theme of task specialization and became famous for his investigations of its impact on manufacturing activities. Among other things, he showed that manifold gains in productivity could be achieved in manufacturing activities if all workers focused their efforts on one simple task of an assembly line (Smith, 1910; 1910a, originally published in 1776). This set the stage for the division of labor seen throughout the Industrial Revolution.

Smith's theories, in turn, influenced Eli Whitney, who pioneered the practical implementation of mass production in the late 1790s in the United States. Whitney and Simeon North applied mass production techniques to the manufacture of guns, for which they secured large multi-year government contracts. At about the same time in England, James Watt and Matthew Robinson Boulton developed the concepts of standard operating procedures, production cells, and incentive payments.

The Industrial Era

His contributions to manufacturing management notwithstanding, James Watt is probably better known for having perfected the steam engine (patent granted in 1769), thus paving the way for the First Industrial Revolution, generally seen as the period from around 1770 to 1850. The First Industrial Revolution produced managerial challenges that were previously unthinkable. New manufacturing techniques drove a tremendous expansion in the markets for machines built around steam engine principles. Steamships and locomotives were developed, which led to a tremendous growth in canal and road transport and, in turn, trade in general. These gave rise to a number of new management problems and situations, as well as opportunities for organizational development.

The two main management figures of the First Industrial Revolution were Henry Fayol and Frederick Winslow Taylor. Fayol pioneered what became known as "functionalism," a set of prescriptions for structuring large organizations around forecasting, planning, and coordination activities. Later, Alfred P. Sloan successfully put functionalism into practice at General Motors.

But the giant of this period was undoubtedly Taylor, whose principles of scientific management (Taylor, 1911) had an impact that extended well beyond his time, an impact that is arguably unrivalled in the history of management thinking.

Taylor, who was born in Germantown, Pennsylvania, was perhaps one of the first actual organizational development consultants. He believed in continuous improvement through the careful and precise measurement of the times and motions involved in relatively simple manufacturing activities, such as moving iron and steel bars from one location to another in a factory. After having worked in several different positions at a company called Midvale Steel Works, and starting up his own new capitalist venture, Taylor took on a management-consulting career with the publication of the best-selling book titled *A Piece Rate System* (Taylor, 1885). In spite of its strong dehumanization element, which led to considerable union opposition, Taylor's system still lives on to some extent, particularly in organizational development approaches focused on increasing productivity in factories (and without a direct concern with quality).

Taylor's antithesis in management conviction terms was Elton Mayo, a social psychologist born in Adelaide, Australia. The significance of Mayo's contributions to organizational development lies in his ideas about the importance of non-economic rewards and personal satisfaction to employee productivity. His criticism of the model proposed by Taylor was rooted in the fact that the model devised optimal work procedures and somewhat imposed them on the workers through a system of simple financial incentives, without giving workers the opportunity to provide their own input. That is, Taylor's sin, according to Mayo, was not to allow those who would have to put into practice the key organizational changes, to participate in the decision-making process that led to the changes.

Mayo provided scientific evidence that taking workers' attitudes and personal motivations into consideration when designing work paid off in economic terms. He investigated the relationships between people working together and, unlike Taylor, paid relatively little attention to such things as procedural routines, times, and motions. What both Taylor and Mayo shared, however, was their main goal, which was the improvement of organizations.

Mayo's most important research project was labeled "The Hawthorne Investigations," a 10-year project conducted at Western Electric Company's Hawthorne Works in Chicago. The project began in 1927 and involved around 20,000 subjects and 100 investigators. Its main focus was on the behavior of small groups under different physical working configurations and social stimuli. Among its main findings was the notion that, for the average worker, the desire to stand well with one's fellows and managers easily outweighs the influence of financial rewards and physical working conditions. This notion became known as the "Hawthorne effect."

The world was well into the Second Industrial Revolution, a period from approximately 1850 to the years preceding the official start of World War II, when Mayo's studies began. This period saw many successful organizational development practitioners, some of which were full-time executives at large companies that later became landmarks in the history of the corporate US Among these are Henry Ford I and Alfred P. Sloan.

Many view Henry Ford I, founder of Ford Motor Company, as the inventor of the automotive assembly line. He is also widely perceived as the first to use mass assembly line production as a means to successfully compete based on price. Alfred P. Sloan, an MIT-trained engineer and former president of General Motors, is credited with the development and practical implementation of the concept of a "multi-division company." He implemented this concept by breaking up General Motors into a set of independent divisions (each with its own engineering, production and sales departments), all of which reported to a corporate control division comprising mostly senior management personnel[1].

The Post-War Era

The Industrial Era was marked by an organizational development focus on autocratic and control-centered approaches, which placed emphasis on managers rather than employees. Even Mayo, who added a more humanistic view to the problem of making organizations more productive, has been strongly criticized by what some view as an objectification of human beings. While he recognized that the workers' organizational beliefs and motivations were important determinants of productivity, some accused him of blindly adhering to the industry's own view of employees as means to be manipulated or adjusted for impersonal ends. Among the most important of such impersonal ends were, of course, business profits, which do not seem so impersonal when one thinks about the kind of lifestyle a good chunk of them bestows on senior management.

After World War II, in the period that I generally refer to here as the Post-War Era (from the end of the war to the late 1980s), some influential management thinkers embraced the management-centered view of organizational development that was prevalent in the Industrial Era. One such thinker was Douglas McGregor, who argued that the basic beliefs held by managers deeply influence the inner workings of organizations. This general hypothesis framed the development of McGregor's hierarchy of needs and his two theories of how management behavior shapes organizations. These theories became known as Theory X, which refers to autocratic and controlling managers, and Theory Y, which refers to managers who are more democratic and willing to delegate responsibilities to workers in a decentralized manner. According to McGregor, these two theories were two extremes of a continuum that were not usually found in organizations in their pure form.

Nevertheless, a former collaborator of McGregor, psychologist Abraham Maslow, became a strong champion of the implementation of Theory Y in organizations, combining it with a refined version of McGregor's hierarchy of needs. By adopting Theory Y, organizations would benefit from what Maslow saw as an inherent characteristic of human beings—a deeply rooted need to work and feel like a valuable member of a society. Although Maslow's idealistic strain has been strongly criticized, it provided a welcome shift from the organizational structure-centered themes that dominated organizational development thinking in the Industrial Era.

A contemporary of Maslow, clinical psychologist turned organizational development consultant Frederick Herzberg, was another nonconformist of this period. He developed a theory of organizational work that separated stimulating factors in the workplace into two main categories—hygiene and motivation factors. According to Herzberg's theory, both types of factors must be satisfied if optimum productivity and quality is to be achieved. Hygiene factors are related to the satisfaction of basic animal needs, such as the need for nourishment and health care. Motivation factors are related to the satisfaction of social and intellectual needs, such as the need to be liked by others.

Another way of looking at Herzberg's theory is that hygiene factors are "dissatisfaction" factors in the sense that if they are not present, dissatisfaction ensues. Motivation factors, on the other hand, are "satisfaction" factors in the sense that they lead to increased job satisfaction once hygiene factors are in place. Herzberg's theories are seen as the basis for many contemporary job enrichment approaches, such as flextime (allowing workers to work according to flexible time schedules) and several different worker benefits and compensation schemes.

The Post-War era also saw the emergence of two of what were later to become very influential approaches for organizational development—action learning and organizational learning. Reg Revans, who pioneered action learning, argued that small groups of peers, from factory floor workers to managers, could learn from and support each other in order to achieve significant gains in productivity and quality. Revans insisted that only those who are directly involved in doing the work can effectively improve it, an idea that is at the root of the development of "quality circles" in Japan (small teams of employees, generally from the same organizational area, who meet regularly to discuss and eliminate quality-related problems).

Chris Argyris, who criticized formal organizational systems stemming from the scientific management movement for neglecting both the social and egotistical needs of individuals, pioneered the concept of organizational learning, an idea that underlies knowledge sharing, one of the main topics of this book. Argyris introduced important concepts, such as those of single-loop and double-loop learning, and worked closely with Donald Schon, a philosopher and business consultant. Single-loop learning is predominantly reactive and attempts to maintain the status quo, no matter how bad it is. Double-loop learning is proactive, aimed at changing the structures and paradigms that underlie well-established yet unproductive work practices. Argyris and Schon published several books together, including *Organizational Learning* in 1978 (Argyris & Schon, 1978).

The Post-War era also provided a nurturing environment for the birth and growth of the quality movement. The two principal figures of this movement were William E. Deming, a statistician with a Ph.D. in physics, and Joseph M. Juran, a former engineer at American Telephone and Telegraph (AT&T).

Deming was one of the first organizational development thinkers to suggest a shift from problems to processes as the focus of organizational development. He argued for business process-focused improvement methods that emphasized the use of statistics. Deming is credited with part of the economic turnaround that happened in Japan from 1950 to 1980. According to him, all business processes are subject to a certain degree of variability, which can reduce the quality of business process outputs. If this degree of variability is reduced, average quality will increase as a result. Deming also extended Pareto's rule to management by arguing that workers cause only 20% of manufacturing problems, while business process design and management systems cause 80% of those problems.

Like Deming, Juran worked as an organizational development consultant in Japan in the years that followed World War II. However, he did take a different approach than Deming, focusing mostly on top management practices for ensuring quality. Juran suggested that Deming was more comfortable with statistics than with management issues, and that what were often seen as visionary statements were little more than far-out statements and platitudes that gained acceptance, as did many other management ideas, by being repeated over and over again.

Still, some believe that Juran's influence on organizational development thinking was a minor one when compared with Deming's, whom many view as the father of what later became known as the total quality management (TQM) movement.

Finally, one more influential movement, whose peak occurred in the 1980s, was the "excellence" movement. The most popular writers of this movement are undoubtedly Tom Peters and Robert Waterman, whose best-seller *In Search of Excellence* sold over five million copies worldwide (Peters & Waterman, 1982). It is difficult to single out a set of major ideas that emerged from Peters and Waterman's early excellence movement, which was seen as a group of disconnected practices by successful companies conveyed in a very simplified (maybe oversimplified) and optimistic way. One of Peters' subsequent books, *Passion for Excellence*, introduced the concept of "management by walking around" (MBWA), which also enjoyed some popularity.

Another contributor to the excellence movement was Rosabeth Moss Kanter, a respected scholar in her own right, whose book *The Change Masters* was hailed as the thinking manager's *In Search for Excellence*. Throughout her life, Kanter has advocated change and innovation in organizations and has developed guidelines for building a change culture in organizations. These guidelines place emphasis on change-focused collaboration at all organizational levels, and on allowing change to occur from the bottom-up.

The Business Process Management Era

Upon comparing the Post-War and the Industrial eras, one can notice a shift in the focus of organizational development approaches between the two periods. In the Industrial Era, emphasis has consistently been placed on the design of optimal procedures and organizational structures by managers, and their enforcement from the top down. In contrast, the Post-War era presents a clear trend toward more participatory management styles and a shift of interest from purely structural to social issues. However, the organizational development ideas in these two eras share one common characteristic. Most of them emerged from academic investigations and hands-on experiences in manufacturing settings.

However, the economic environment in most countries has been witnessing rapid changes since the mid 1970s, particularly in the developed and developing nations. In the United States, for example, the number of white-collar workers surpassed that of blue-collar workers around 1976, which suggests to some extent that work in general has gradually become more knowledge intensive, requiring better educated workers. Estimates by the Organization for Economic Cooperation and Development (OECD) indicate that, since the 1990s, for every three dollars spent in the United States and several other developed countries, around two dollars have been spent on services. A good percentage of service sector sales come from companies that sell some form of data product (e.g., computer software, financial indexes, and news in general).

The above macroeconomic trends have similarly been observed in individual organizations. Some of my own studies, conducted in collaboration with other researchers, have shown that over 70% of exchanges within organizations involve information or knowledge (e.g., paper and electronic forms, memos, faxes, and e-mail), as opposed to tangible things like parts, raw materials, or tools (Kock & McQueen, 1996; Kock et al., 1997; Kock, 2003). This is true even when the objects of the analysis are manufacturing organizations. In service organizations, the proportion of exchanges involving information or knowledge can be over 95%, with exchanges of tangible items making up 5% or less of all organizational exchanges.

Thus, it is no wonder that one of the main organizational development movements in the 1990s, the business process reengineering movement, has focused heavily on service activities from its inception in the early 1990s. The emergence of that movement marks the beginning of what I refer to here as the Business Process Management Era, which is characterized by a great concern in organizations with the improvement of productivity and quality of business processes, as well as the implementation of business process-centered organizational strategies, such as business process outsourcing and organization-wide business process integration through the use of enterprise systems.

Business process outsourcing entails the farming out of entire processes to external suppliers, building heavily on the infrastructure provided by the Internet. As the name implies, business process integration entails the electronic integration of several organizational processes involved in a company's supply chain, from ordering, passing through production, invoicing, and inventory control to distribution and/or delivery. Key to business process integration is the use of large enterprise systems such as the market leader in this segment, SAP/R3, which has been developed and commercialized by SAP, a German software development company and one of the largest such companies in the world.

The reengineering movement emerged from the work of two consultants, both with solid academic backgrounds—Michael Hammer (who worked in collaboration with James Champy) and Thomas Davenport (who worked in collaboration with James Short). Unlike some of their organizational development predecessors in the Industrial and Post-War eras, Hammer and Davenport built on the work of practitioners (mostly executives from large corporations) to develop the idea of reengineering. Among other companies, they studied how Ford had dramatically improved its accounts payable process, IBM Credit its financing quotation process, and Mutual Benefit Life its new policy writing and approval process. None of those improvements was originally done under the reengineering flag. Nevertheless, due to their underlying similarities, all were later reported as successful cases of business process reengineering.

The term *business process reengineering* refers to radical organizational redesign projects, particularly when they are focused on cross-departmental business processes or sets of interrelated activities. In essence, a business process is like a cake recipe. It has a set of activities that bear some kind of dependence on each other (e.g., the outputs of some are inputs to others) and that must be carried out in a certain order using certain raw materials and tools, and whose final output is a product that is going to be consumed or used by someone (e.g., a chocolate cake). In an organizational context, a business process can be understood as a set of interrelated activities, usually carried out by teams, whose outputs are the goods or services that are typically sold by an organization to its customers (the business process concept is discussed in more detail later in this book).

Reengineering emerged as a reaction to the TQM movement, which in turn was largely based on Deming and Juran's ideas, mentioned earlier. According to Hammer and Champy, TQM projects usually led only to 5-15% improvements in process productivity, while reengineering could lead to improvements of as much as 300%. They also argued that, in spite of TQM's history of success in the US, TQM was better tailored to the Japanese than to the American culture. According to them, the radical approach taken by reengineering fit the American ingenuity and creativity much better than the incremental "small-minded" approach taken by TQM. Moreover, radical business process redesign, as proposed by reengineering, could only happen with the creative use of information technology, which was well in tune with the explosion of the use of networked computers seen in the 1980s and 1990s.

Reengineering went from an idea in the early 1990s to a US $50 billion management consulting industry around 1995. By then, a large number of reported cases of reengineering failure (estimated at around 70% of all reengineering attempts) were pushing the movement into a passing fad status. Among the explanations for such a high failure rate was that reengineering had clearly borrowed some of the methods that prevented Taylor's Scientific Management from succeeding with well-educated workers. For example, it placed radical business process redesign decisions largely in the hands of top managers, key employees, and consultants, leaving those who executed the processes practically out of the business process redesign decision-making loop. The term *reengineering* was soon linked to corporate downsizing and massive job cutting, and thus faced strong opposition from workers and their unions.

Analogous to what happened at the end of the Industrial Era, organizational development approaches in the late 1990s were reverting away from reengineering and leaning towards "softer" approaches that emphasized management/worker collaboration, decentralized access to information and knowledge, and delegation. One such approach is that of organization learning, as proposed by MIT Professor Peter Senge (1990), which focuses on organizational knowledge building and sharing to support optimal teamwork. This approach is essentially a revival of the homonymous approach originally proposed by Argyris, with the basic difference that it takes on a very "soft" and somewhat "evangelical" tone. It appropriately points out the importance of knowledge sharing, but at points adopts a sort of lets-all-be-friends-and-everything-will-be-great view of organizations.

Along with a softer orientation, organizational development in the late 1990s has seen an increasing interest in knowledge[2] management and sharing approaches. For example, Davenport, one of the main figures of the reengineering movement, openly acknowledges reengineering as "something of the past" and has concentrated his efforts on the study and implementation of effective knowledge management techniques in the mid and late 1990s.

However, the now famous Internet stock bubble burst in early 2000 changed the organizational development climate substantially and gave reengineering a new boost. Toward the end of the 1990s, many companies, whose business relied heavily on the Internet or on other Internet-based companies, saw their market values skyrocket, only to see those values take an unprecedented nosedive in early 2000. For example, Yahoo, an Internet portal company whose revenues come primarily from advertising on its Web sites, saw its stock price increase over 1,000% from 1998 to early 2000, only to have it go down to 1998 levels in late 2001.

The Internet bubble burst was followed by a series of difficult events, the most important of which were the infamous terrorist attacks against the World Trade Center and the Pentagon. Those events deeply affected the US economy, as well as the rest of the world, creating a climate of economic instability leading to massive layoffs and desperate attempts by organizations to cut costs and improve productivity. That proved to be a fertile ground for reengineering ideas to gain renewed acceptance, although often under different banners.

An Historic View of E-Collaboration

As with the history of organizational development, the history of e-collaborative computing can be split into a few distinct chronological phases. Even though the widespread use of e-collaboration technologies is closely related to the implementation of local and wide area networks, particularly the Internet, a few precocious projects were developed based on early mainframes.

The history of e-collaborative computing can be organized around four main phases: (1) the Mainframe Era; (2) the establishment of computer networks; (3) the expansion of local area networks; and (4) the Internet Era. Each of these phases is marked by the dominance of a particular computer technology and related attempts to develop e-collaboration technologies. The remainder of this section briefly discusses each of these phases individually.

The Mainframe Era

As its name implies, the Mainframe Era was marked by a dominance of large computer systems, usually known as *mainframes*. It extended from the early 1950s, with the emergence of the first mainframe assembly lines, to the late 1960s, with the first major computer networking projects. A growing presence of computers in organizations and an almost complete lack of concern about e-collaboration distinguish this phase.

Typical mainframe configurations involve a large central computer (the mainframe) connected to a number of "dumb" terminals (i.e., terminals with very limited or no processing capacity of their own—hence the term *dumb*). Unlike the server-client local and wide area network configurations seen today, where client workstations (e.g., personal computers connected to an office network) are equipped with powerful processors, mainframe terminals were used almost exclusively as input/output devices.

In the Mainframe Era, computers were used primarily for their data processing power, rather than their potential for supporting communication and collaboration among groups of workers. Nevertheless, such e-collaboration support ability was timidly explored around the end of this period as new operating systems with rudimentary (compared with what is available today) synchronous and asynchronous collaborative features were developed[3]. One example of such early collaborative operating systems is Multics, developed at the Massachusetts Institute of Technology and first installed in 1967 on a General Electric GE-635 mainframe.

Although terminals allowed decentralized access to information, this was prevented by a high hourly cost of mainframes. For example, the cost of an IBM-7094 in 1966 was approximately $2 million, while its life span was approximately five years. This meant that the hourly cost of the mainframe was about $45, at a time when workers earned as little as $1 per hour. Because of that, and in an obvious mismatch with the organizational development ideas of this time, information was extremely centralized in the Mainframe Era.

Employees of central data processing departments (sometimes referred to by the acronym CPDs) usually took care of all of the data storage and processing activities of organizations. This meant that requests for vital information to perform organizational activities (such as customer contact information) necessarily had to be sent to and fulfilled by CPDs. This situation persisted for many years and was dramatically changed with the advent and expansion of the local area networks. This occurred only after the first computer networks (initially large and geographically distributed ones) were set in place, mostly through government-funded projects in the US.

The Establishment of Computer Networks

This phase ranges from the late 1960s to the mid-1980s. It began with a major development in 1967 (the official start date according to most accounts), the ARPANET project, which provided the basis on which the now ubiquitous Internet has evolved. The ARPANET project began under the auspices of the Advanced Research Projects Agency (ARPA), a branch of the US Department of Defense (DOD). Its main goal was to build a network of shared computational resources by interconnecting major universities and research centers in the US.

A major limitation of early mainframes and their operating systems was their lack of interoperability. In 1966, mainframes of a certain brand could only exchange data with other mainframes of the same brand. International Business Machines (IBM) mainframes could interact only with other IBM mainframes; Burroughs interacted only with Burroughs; General Electric with General Electric; and so on. One of the ARPANET project's main goals was to put together a heterogeneous network connecting IBMs, GEs, and all other mainframe brands together. Another goal was to build a network of dispersed and powerful computer resources so that, in the event of a war-related attack from the Soviet Union[4], only part of the United States' computing power would be actually affected.

At the same time as the ARPANET project began, smaller mainframes were developed, some of which had processing powers similar to earlier mainframes. This was enabled by the development of integrated circuits, which combined many transistors into a single dedicated chip. Such small mainframes were labeled minicomputers. As integrated circuits became smaller and smaller, and the number of transistors that were combined into a chip increased, minicomputers later evolved into microcomputers (aka personal computers or PCs) with ever-expanding processing power and storage capacity.

One of ARPANET's component systems was an early and rudimentary version of e-mail. Given little importance in the beginning, that rudimentary e-mail system rapidly became one of the most widely used components of the ARPANET (Sproull & Kiesler, 1991). By 1971, there were about 37 hosts connected to the ARPANET network, each serving many dumb terminals. Virtually all users of these terminals could now communicate with each other through e-mail, and so they did. Researchers, university professors, and students began using the system for both formal and information communication. Physically dispersed research groups emerged, pushing ARPANET beyond the United States to England and Norway in 1973. The seed of the modern e-collaboration technologies had been planted.

At the same time, pioneering implementations of computer systems to support collaborative work occurred elsewhere, mostly using asynchronous technologies in very experimental ways. These early implementations led to the first organized workshop on computer-supported cooperative work (CSCW), a term that later became synonymous with others such as groupware, computer-supported collaboration, and collaborative computing (Grudin, 1994a). This workshop, conducted in 1984, was followed two years later by the first international conference on CSCW in Austin, Texas (Bannon, 1993).

Several years later in 2001, Kock and colleagues used the term *e-collaboration* for the first time when they guest-edited the first special issue of a journal addressing the topic (Kock et al., 2001). The special issue was published in the *Journal of Systems and Information Technology*, a scholarly publication from the Department of Information Systems at Edith Cowan University in Australia, edited by Craig Standing.

The Expansion of Local Area Networks

The local area networks (LAN) phase goes from the mid-1980s to the early 1990s, and owes much of its existence to the development and widespread use of personal computers. With the development in the mid-1970s of large-scale integrated circuits, whose transistor capacity was much higher than the normal integrated circuits, computers became smaller in size and more powerful. As a result, the first PCs were developed. PCs (also referred to as microcomputers or micros) were usually much cheaper than their predecessors, the minicomputers.

The first of all PCs is believed to have been the Altair, a $400 computer developed in 1974 based on Intel Corporation's 8080 microprocessor. Microsoft founders Bill Gates and Paul Allen developed a simple programming interface and code interpreter for the Altair computer, based on the BASIC programming language. At that time, Allen was a college dropout working for Honeywell, and Gates was a freshman at Harvard. BASIC, which stands for Beginner's All-Purpose Symbolic Instruction Code, was one of the first popular computer programming languages.

The development of Altair was followed by the release of increasingly more sophisticated PCs, which relied on new central processing unit (CPU) technologies developed by Intel, Motorola, and other CPU manufacturers. In 1978, Intel released its 8086 processor, which contained over 29,000 transistors. IBM used the Intel 8088 (released in 1979) in their famous and much cloned IBM PC—hence the later widely used term "IBM PC-compatible" to refer to standard PC hardware and software.

As microcomputers grew cheaper and more powerful, established software developers started shifting their efforts toward popular PC platforms. Countless new software developers entered the PC application development market, taking advantage of lower entry barriers stemming from the proliferation of suppliers of basic PC components. With the development of micro-mainframe connection cards, such as the once widely used IRMA cards, PCs could be used both as independent computers and dumb terminals. Due to their newly obtained dual functionally, PCs became an attractive alternative to dumb terminals used with mainframes. This paved the path for the downsizing[5] of applications from mainframes to local area PC networks, whenever possible. Applications that required light processing power, low storage capacity, or that could benefit from distributed processing arrangements, became the object of massive downsizing.

Once-very-wealthy companies that relied heavily on mainframe sales for financial growth saw their market shrink and revenues dwindle. At the same time, PC and LAN companies experienced a tremendous growth. As the PC manufacturing industry became more competitive and PC prices plummeted, the importance of software that linked several PCs and resources (e.g., printers, fax machines) increased. The market for such software systems, generically referred to as local area network operating systems, skyrocketed. Many companies tried to enter this market, but its undisputed leader in the 1980s was Novell Corporation, a Utah-based company that launched its first LAN operating system in 1983.

LANs emerged as a standard tool for organizational development in firms of all sizes. At the beginning, one of the main advantages of having a LAN connecting several PCs and computer peripherals was that it enabled the sharing of what were then relatively expensive resources, such as laser printers and large-capacity/high-speed hard disks. In time, however, LANs also progressed as a mechanism to implement and consolidate some of the organizational development approaches that became popular in the 1980s. At this time, the total quality management and the excellence movements called for, among other things, worker empowerment through delegation and decentralized information access, for which many viewed LANs as the ideal support tool.

Many synchronous and asynchronous e-collaboration technologies were developed in this phase. Some of them, such as Information Lenz and The Coordinator, extended the common set of features exhibited by early e-mail systems. Others, such as GroupSystems, Teamfocus, and MeetingWorks, provided support for decision-oriented face-to-face meetings. Still others, such as Lotus Notes and Domino, operated as suites on which customized e-collaboration technologies could be developed to support specific group processes. Most of these technologies resulted from the work of researchers in universities and government research centers, as well as corporate research centers. The following phase, the Internet Era, has seen many of these applications migrate from LANs to Internet-based platforms.

The Internet Era

The Internet Era began in the early 1990s and extends to the present day. Its emergence is due largely to technical limitations of LANs regarding remote communication and data access. Similar to isolated PCs, isolated LANs need special devices and communication media to share data over long distances. While LANs can link PCs in the same building or campus, without wide area networks such as ARPANET, PCs in different LANs are unable to exchange data among themselves effectively. A rather ineffective approach to make those PCs use data stored on each other's hard disks, often used before the 1990s, is to swap disks with data among PCs in different LANs, using the postal system to mail disks among different locations. Links between LANs should be built for those data exchanges to happen more effectively.

The service sector growth seen in this period, and particularly the growth in information industries (e.g., software development, TV news, newspapers, and financial information firms), contributed to a huge increase in the amount of data that had to be handled on a daily basis by organizations in general. Besides, this period was marked by a fast pace of change, which called for radical change approaches to organizational development such as business process reengineering. Such a fast pace of change also required that the data sitting in isolated LANs be moved around faster than ever before, especially in the increasing number of companies spanning city, state, and even national boundaries.

One of the first successful attempts to link computers separated by large distances was the already mentioned ARPANET project. However, the main goal of this project was to link mainframe computers, not PCs already connected in LANs. In time, basic devices (which included their own hardware and software, and sometimes ran on PCs) called *bridges* were developed, to link two or more remotely located LANs. A bridge can use different media to connect two or more LANs. Many of the early bridges used private telephone lines (copper wires) with speeds only up to 64 kilobits per second for inter-LAN connectivity (i.e., a single 10-megabyte file would take over 20 minutes to be transferred between LANs). These speeds increased as new data compression techniques, thicker copper wires with increased bandwidth, and new communication media (e.g., fiber optics) became commercially available.

However, connecting LANs through bridges was expensive and complex, particularly when several LANs had to be interconnected. If users connected to a LAN at one company wanted to access data located in another LAN at a different company, then at least one dedicated bridge between these two LANs would have to be set up and administered. What was really needed was something (primarily a technical solution) simpler and less expensive than that. Perhaps a single public infrastructure that allowed multiple LANs to be linked, regardless of the existing hardware and operating system configurations, would do the trick. With this in mind, Tim Berners-Lee proposed in 1989 a global hypertext project, later known as the World Wide Web (WWW or the Web). The implementation of this project started in October 1990 and led to the development of the first Web server and browser, early specifications of the HTTP communication protocol, and the HTML language.

Contrary to a popular perception, the Web is not the Internet. The Web is an abstract collection of sites created by Web servers (e.g., the server that runs Amazon.com's site) that uses the physical infrastructure provided by the Internet. The Internet, in turn, evolved from the initial infrastructure set in place for the ARPANET project, which was, as mentioned before, commissioned by the US DOD. In 1987, the National Science Foundation took over its administration from the U.S. DOD. The Internet was not very popular then or in the years immediately following Berners-Lee's project. It really started to show signs of exponential growth in 1993, when the popular Web browser Mosaic was developed at the University of Illinois, Urbana-Champaign. This landmark development formed the basis for the emergence of the Web, as we know it today.

Since the beginning of the Internet, there was a variety of Internet hosts, which included file transfer protocol (FTP) hosts allowing file uploads and downloads. Web hosts (or Web servers), used together with Web browsers such as Mosaic, provided easy access to files by means of hyperlinks, which users needed only to click on to download files. The total number of Internet hosts was 1 million in 1993, 2 million in 1994, 5 million in 1995, 10 million in 1996, and so on, reaching a whopping 147 million in 2002.

The number of Web hosts grew at a similar rate. In 1993, the Web had just 130 Web hosts[6]. In 1994, it reached about 600, growing to approximately 20,000 in 1995, 100,000 in 1996, 1.8 million in 1998, 4 million in 1999, and 36 million in 2002. Popular commercial Web browsers succeeded Mosaic, notably Netscape Navigator (developed by members of the original Mosaic team) and Microsoft's Internet Explorer. Such browsers are used today by hundreds of millions of users around the world to obtain information, purchase goods and services, and build personal relationships over the Internet.

The advent of the Internet Era led e-collaboration technologies to move from the LAN environment to the Internet, mostly as client-server systems running on platforms made of generic, platform-independent Web browsers (on the client side), and platform-dependent Web servers (on the server side). For example, Internet Explorer, which is a Web browser, runs on many platforms, from several versions of Windows to the Mac operating system. Microsoft's Internet Information Services, which is a Web server, runs primarily on the Windows servers.

Interfaces to popular e-mail packages, such as Groupwise by Novell Corporation, were developed so users could access their mailboxes using any standard Web browser. In the mid-1990s, companies like Qualcomm and

Microsoft offered free Web-based e-mail, available to anyone with access to a standard Web browser. Synchronous "chat" tools were developed to run on Web pages, which allowed for simultaneous two-way transfer of voice and text, threatening to bite into the profitable long-distance telephone market. Video and audio streaming allowed for real-time video and audio transfer and broadcasting, even through narrow bandwidth media such as copper wires.

The Web browser interface has become, in the Internet Era, the standard for e-collaboration. The low-cost public infrastructure provided by the Internet brought about seamless integration of data and resources to what was formerly a chaotic mix of dispersed LANs, most of them isolated from other LANs. In the Internet Era, the problem is no longer accessing information located elsewhere, but coping with information overload.

Summary and Concluding Remarks

From a historic perspective, organizational development has progressed hand-in-hand with new management ideas. At the same time, the success of computer technologies in general, and e-collaboration technologies in particular, have depended in part on the adequacy with which they were able to support popular organizational development ideas of the time. This explains, to some extent, the proliferation of collaborative "knowledge management" computer tools in the late 1990s, which occured at a time when there was heightened interest in organizational development approaches that promoted cross-functional knowledge sharing.

The evolution of organizational development can be split into three main phases—the Industrial Era, the Post-War Era, and the current phase. The Industrial Era was marked by an emphasis on the design of optimal procedures and organizational structures by managers, and their enforcement down the organizational ladder.

In the Post-War Era, the emphasis shifted from purely structural to social issues, and top-down control was replaced with participatory management. While essentially different, the Industrial and Post-War eras were primarily concerned with manufacturing activities and organizations. This started to change in the late 1980s with the advent of organizational development ideas targeted at the service sector, where most knowledge-intensive industries are now located.

Analogous to the organizational development evolution, a historical look at the evolution of e-collaboration technologies can be divided into four main phases—the Mainframe Era, the establishment of computer networks, the expansion of local area networks, and the Internet Era.

In the Mainframe Era, which was marked by the hegemony of large mainframe computers, e-collaboration technologies were practically nonexistent. In this phase, which went from the early 1950s to approximately 1965, e-collaboration technologies were fairly rudimentary and confined to running on top of mainframe operating systems.

The phase where the establishment of computer networks occurred ranged from the late 1960s to mid-1980s. This phase was marked by the development of the ARPANET project, which began in 1967 under the auspices of the Advanced Research Projects Agency (ARPA), a branch of the US DOD. The main goal of the ARPANET project was to build a network of shared computational resources by interconnecting major universities and research centers in the US. A few initially neglected component systems of the ARPANET network, particularly its simple e-mail tool, planted the seed for the future development of more sophisticated e-collaboration technologies.

From the mid-1980s to the early 1990s, the organizational world has seen the expansion of LANs, which connected personal computers. This phase owes much of its existence to the development and widespread use of powerful and reasonably cheap microcomputers (personal computers), and was the most prolific in terms of the development of new e-collaboration technologies.

Most of the e-collaboration technologies made available in the phase that followed the expansion of LANs (the Internet Era) are based on technologies developed in the preceding phase. The Internet Era extends from the early 1990s to the present day. Its emergence is largely due to remote communication and data access limitations of LANs, which supported the interconnection of microcomputers located relatively close to each other (i.e., in the same building or campus). The advent of the Internet Era led e-collaboration technologies to move from the LAN environment to the Internet, mostly as client-server systems running on platforms made of generic, platform-independent Web browsers (on the client side), and platform-dependent Web servers (on the server side).

In the Internet Era, the Web browser interface has become the primary standard for computer-mediated collaboration. The low-cost public infrastructure provided by the Internet brought about seamless integration of data and resources to what was formerly a chaotic mix of separate LANs. In the Internet

Era, the problem of accessing remotely located information has been solved. The real challenge in this phase is to deal with the explosion of information available at our fingertips.

Endnotes

1 More recently, such corporate control companies have often been referred to as *holding companies*.

2 The concepts of data, information, and knowledge are defined and contrasted later in this book.

3 *Synchronous* group e-collaboration requires same-time communication (e.g., a teleconference), whereas *asynchronous* e-collaboration usually involves time-disconnected communication (e.g., the use of e-mail for work-related communication).

4 The Cold War between the US and former ally, Soviet Union, was well under way by this time.

5 The term *downsizing* has different meanings, depending on the context in which it is used. Popular business magazines often use it to refer to the reduction in the size of organizations, often caused by massive layoffs and division sellouts. In information systems circles, the term is often used to refer to the migration of computer applications from mainframes to local area networks.

6 Web hosts store Web sites, which, in turn, store text, video, audio, and pictures that can be downloaded by Web browsers (e.g., Mosaic, Netscape Navigator, and Internet Explorer).

Chapter III

What is a Business Process?

The Pervasiveness of Business Process Improvement

Business process improvement can be defined as the analysis, redesign, and subsequent change of organizational processes to achieve performance and competitiveness gains. The idea that business process-focused improvement can be used as a tool to boost organizational performance and competitiveness is not new. In fact, business process improvement has been the basis of several widely adopted management approaches, such as total quality management, business process reengineering, and organizational learning. As the following sections briefly show, business process-focused improvement can be a unifying concept of these management approaches.

Total Quality Management

One of the main tenets of the total quality management movement is that the focus of improvement should be on *processes* rather than *problems*. Moreover, when one carefully looks at the causes of organizational problems (e.g., high costs, low quality, deficient worker productivity), a variation of the famous "Pareto rule" reflects pretty well what actually happens. That is, 80% of those problems are business process-related, whereas only 20% can be blamed on the people who carry out business process activities (Deming, 1986).

This new perspective constitutes a shift from the old view that problems in organizations are caused by workers' negligence and their disregard of management-set rules for business process execution. Organizational norms of accepted behavior, formal job definitions, rigidly set communication channels, hierarchical structures, inflexible computer systems, as well as reward systems are just a few elements of organizational process design. It is in these elements, argued the total quality management movement, that most improvement opportunities can be found.

When William E. Deming, the main figure of the total quality management movement, unequivocally proposed these ideas, many thought that he was either wrong or, on the opposite extreme, pointing out things that were too obvious to be relevant. As it became clear in the 1980s, especially to the American business establishment, not only was he right, but, given the major changes that his ideas generated in many businesses, he was also proposing ideas that proved to be fairly counterintuitive at the time. Deming's story is one of the most successful in the whole history of management thought.

Business Process Reengineering

Unlike the total quality management movement, which seems to have been built around a common set of concepts and ideas, the business process reengineering movement has been characterized by the existence of different schools of thought. From the beginning, at least two schools of thought could be identified.

Computer expert and management consultant Michael Hammer led the radical and more popular school of thought in connection with business process reengineering. Hammer and colleagues proposed reengineering as a totally new and revolutionary approach for business process improvement, and argued for a complete departure from the incremental business process improvement approach, which characterized the total quality management (Hammer, 1990; Hammer & Champy, 1993).

The other school of reengineering thought, led by then University of Texas Professor Thomas Davenport, was more conservative in its expectations. It proposed radical business process redesign as just a new tool for business process improvement, especially for those organizational processes that were at the core competencies of an organization (Davenport, 1993; Davenport &

Short, 1990), and argued for the combined use of this new business process improvement tool with others, such as total quality management (Davenport, 1993a).

The radical reengineering school pioneered by Hammer and colleagues was initially much more popular than the conservative school, which is exemplified by the fact that Hammer and Champy's (1993) book, *Reengineering the Corporation*, sold over 2 million copies by 1995 and is still a top-selling book to this day. Davenport's (1993) book, *Process Innovation*, sold less than 100,000 copies in the same period—a sizeable figure, yet only a small fraction of Hammer and Champy's book sales.

However, two years after the publication of Hammer and Champy's (1993) book, the radical school was split. Due to a high rate of failure in reengineering projects (often reported as 70-75%), the radical tone of this school was slightly softened in Hammer and Stanton's (1995) sequel titled *The Reengineering Revolution*, though its authors persistently clung on to the idea that reengineering was a revolutionary approach.

At the same time, a new and short-lived reengineering branch was begun by one of its forerunners, then CSC Index's head James Champy, focusing on shifting management paradigms, as opposed to simply redesigning business processes (Champy, 1995).

Champy argued that the high rate of failure in reengineering projects was, to a large extent, caused by the fact that business processes were being radically redesigned, while the way managers viewed their organizations was not. According to Champy, this dichotomy had led to a widespread lack of management support of reengineering attempts, which was, indeed, found to be one of the foremost reasons why reengineering projects had been reported to fail, according to a study by Caldwell (1994). Nevertheless, the arguments put forth by Champy have been found to be less than convincing and somewhat "evangelical" (Wensley, 1996), and the impact of his 1995 book was small among the torrent of reengineering books and articles published in the same year.

Both the radical and the more conservative schools led by Hammer and Davenport recently converged on one main point—that the main contribution of the business process reengineering movement was not to propose a radical and revolutionary approach for organizational change. Its main contribution has been increased management awareness about the importance of business processes and the advantages of a business process focus in efforts aimed at attaining optimal organizational design and performance improvement (Davenport et al., 1996; Hammer, 1996; Kock & Murphy, 2001).

Organizational Learning

Business process improvement also seems at the core of what is often referred to as the organizational learning movement. Peter Senge, an MIT professor and author of the seminal book *The Fifth Discipline*, is believed by many to be the father of the organizational learning movement (Senge, 1990). Nevertheless, Senge's work builds largely on ideas in the general field of system dynamics, developed a decade earlier by University of Lancaster professor Peter Checkland (Checkland, 1981; Checkland & Scholes, 1990), and by Harvard professor Chris Argyris.

One particularly important contribution by Argyris was the concept of double-loop learning, which he contrasts with that of single-loop learning (Argyris, 1977, 1992). Argyris often seems to adopt a cybernetic perspective (i.e., a control-centered perspective) in his interpretation of learning, hence his choice of explaining the concepts of single and double-loop learning by often using analogies with control devices such as a thermostat-activated heater (Argyris, 1977).

A device that senses a decrease in the temperature of a room and changes the amount of heat that it gives off into the room, in order to maintain a constant room temperature, engages in what Argyris refers to as single-loop learning. This is a reactive behavior that involves little knowledge and that, therefore, can be easily automated. It does not involve the understanding of the mechanics of heat transfer and fluid dynamics.

Double-loop learning, on the other hand, is the type of learning involved in understanding the process through which variations of the room's temperature occur, and devising a more efficient approach to maintain a constant room temperature. Single-loop learning alone would never lead to the identification of an air leak close to one of the windows, for example. Thus, it would lead to a situation in which the room's temperature would be kept constant at a higher cost than it could have been otherwise (assuming that the higher electricity bill would be more expensive than fixing the leak).

In this example, the identification of the leak could only result from an analysis of the heat transfer *process* in the room. The understanding of the mechanics of air heating and flow in different points of the room would eventually reveal that there had been a disproportionate amount of heat being transferred near the window. This is one example of the similarity between what is called learning in the organizational learning movement, and business process-focused improvement. I will extend this discussion later in this book, but now I will answer the question with which I started this chapter: What is a business process?

What is a Business Process?: Different Views

As a concept becomes more abstract, so does the discrepancy in the ways different people construe the concept. A concept that refers to a tangible object like a chair, for example, is likely to be understood more or less in the same way by two people. If one person says, "Then I sat on this big, soft, blue chair," then the other will probably be able to visualize the scene in more or less the same way as the first person did.

With abstract concepts such as a business process, however, this shared mental visualization is much less likely to be achieved without substantially more effort and clarification. One of the reasons for this difficulty is that abstractions are not perceived by our five senses as real objects like a chair is (e.g., we can see and touch a chair), and therefore must be understood based on abstract models. If these models do not exist, or are too rough and incomplete, then there is often a sense of perplexity about what they mean. This phenomenon is an example of what has been called "dissociation of sentience and knowledge," which is discussed in length by Harvard professor Shoshana Zuboff in her landmark book, *In the Age of the Smart Machine* (Zuboff, 1988, 1996).

The disociation between sensorial stimuli and understanding has been associated with adaptation problems faced by workers who are taken away from the shop floor, where they have direct and often physical contact with the machines they operate, and into computer-operated control rooms. In the control rooms, workers must understand the machines they control as points in a production process, which is itself an abstract entity. This may not be easy for factory workers, as suggested by the following quote from one of Zuboff's interviewees.

It is very different now.... It is hard to get used to not being out there with the process. I miss it a lot. I miss being able to see it. You can see when the pulp runs over a vat. You know what's happening.

(Zuboff, 1996, p. 197)

Business processes, like most abstract entities, need to be modeled in some way in order to be understood. And, more importantly, two or more people must understand the business processes in roughly the same way, if they can even begin to improve it. Models, however, irrespective of how complex they are, in most cases are limited representations of whatever they are supposed to depict, whether those are real objects or abstract entities.

A given representation of a transistor, for example, can help one predict how it will be "fired" (i.e., amplify an electrical input) when an electrical impulse of a certain voltage is applied to it. However, the same representation can be almost useless to predict the operation of the same transistor when its input is an alternate current whose frequency is above a certain level, as in discreet analog telecommunication circuits. Similarly, a given representation of a car, such as an owner's manual's diagram explaining the basic operation of the car, can be detailed enough for someone who wants to *drive* the car, and yet useless to someone who needs to *repair* the car. In fact, perhaps the only characteristic that is shared by all models is that they are all incomplete representations of what they attempt to depict.

A few main types of business process models or views are discussed in the following sub-sections. As already discussed, these views lead to incomplete representations of business processes, and, therefore, should be understood in terms of their pros and cons in today's information and knowledge-intensive organizational environments. I try to contribute to this understanding in my discussion of each of the views.

The Workflow View

Although there seems to be little agreement on what a business process is or on what the main elements are that make it up, the predominant view among academics and practitioners seems to be a set of interrelated activities (Hunt, 1996; Kock & Murphy, 2001; Ould, 1995). In this sense, business processes are seen as activity flows (or workflows) composed of activities that bear some sort of relationship to each other (Kock & McQueen, 1996; White & Fischer, 1994). Thus, if activities are not perceived as interrelated, then they are not part of the same business process.

While this is an interesting conceptualization of business processes, there is a catch. Business processes are not real entities. That is, sets of interrelated activities are just mental abstractions that allow us to understand organizations and how they operate. Therefore, business processes are what we perceive them to be—what our mental models tell us about them. For example, if a manager perceives "design a new product" and "market a new product" as being closely related, then the manager will draw a workflow model of the business process "launch a new product" with activities in both the product design and marketing departments. Such a model would be a cross-departmental business process. However, if the manager sees these activities as separate and independent, then his or her view of the business process either will be incomplete or will emphasize the role of one of the two departments (Vennix, 1996).

The difficulty here is a direct result of the fact that business processes are mental abstractions (together with our mental representations of everything else in the world, I must add), and of the "worst" type—mental abstractions of abstract entities, since business processes are not "real" tangible things. Unlike tangible things such as a chair or a desk, business processes are essentially a product of our imagination. They cannot be seen or touched. This characteristic will affect most of what I say in this book about business process improvement and knowledge sharing.

Different people in the field of business process management understand the terms "interrelated activities" and "workflow" differently. This can be an obstacle to achieving a basic goal of any workflow representation, which is the identification of the types of relationships between activities. In fact, such identification of types of relationships has not been a widespread concern, with a few exceptions (see Malone and Crowston's [1994] discussion about workflow coordination, and Schmidt's [1994] conceptualization of business processes involving coordination of different functions). I believe that a categorization of basic activity relationships in business processes is helpful in identifying organizational processes from a workflow perspective.

Figure 3.1: Materials receipt process of a chimney manufacturer (Adapted from Kock et al., 1997, p. 72)

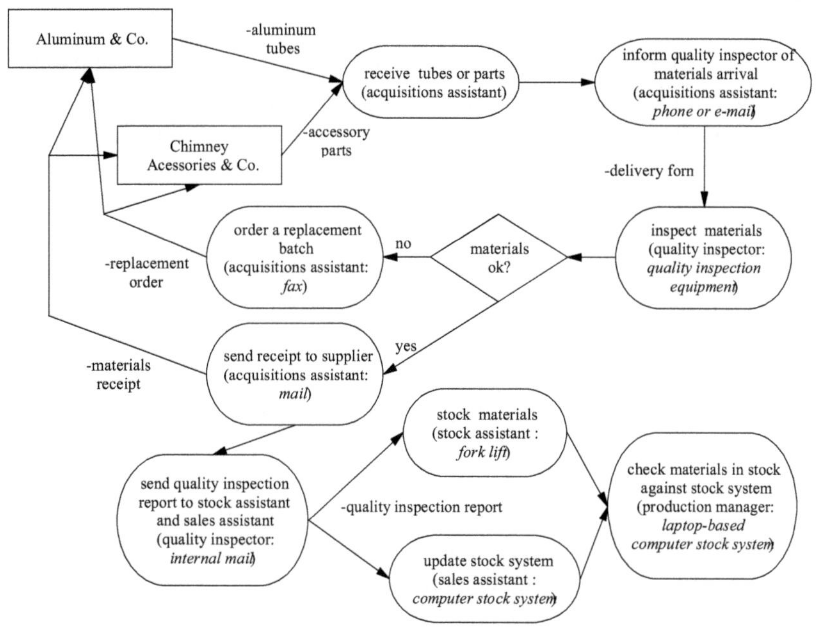

My own study and consulting experience suggest that there are at least three main types of relationships among activities in business processes, which I refer to as (a) common predecessor, (b) common successor, and (c) predecessor-successor. These relationships are illustrated in the "receive materials" business process of a chimney manufacturer shown in Figure 3.1, where activities are depicted in oval shapes, and the arrows indicate the flow of execution of the activities in the business process.

The symbols used to represent business processes as workflows can vary widely, even when professional organization standards are consulted. In Figure 3.1, a rectangular shape represents an external supplier of the business process, whereas a diamond shape indicates a decision point in the business process. Each activity is described by its name, followed (in parentheses) by the organizational function that carries out the activity and the italicized name of the main tool used by this function. Freestanding text beginning with a dash is used to describe a product (i.e., a piece of data or a material thing) that flows between activities.

The *common predecessor* relationship joins together activities that have a common immediate predecessor activity. In the business process shown in Figure 3.1, the activities "order a replacement batch" (carried out by the acquisitions assistant usually by fax) and "send a receipt to supplier" (also carried out by the acquisitions assistant, typically using ordinary mail) display this type of relationship. Both activities have the same immediate predecessor—the activity "inspect materials"—conducted by the quality inspector using specialized quality inspection equipment. This common predecessor must be carried out before each of these two interrelated activities.

The *common successor* relationship connects activities that have a common immediate successor activity. The activities "stock materials" and "update stock system" (the former carried out by the stock assistant with the use of a forklift and the latter by the sales assistant on a computerized stock system) are connected through a *common successor* relationship. Both activities have a common successor—"check materials in stock against stock system"—carried out by the production manager by walking through the stock warehouse and comparing it with the inventory database using a laptop-based version of a computerized stock system.

The *predecessor-successor* relationship, the most common type of relationship between activities, refers to any two activities that take place in sequence, one after the other. Note that, as with the two types of relationships described above, a *predecessor-successor* relationship can exist even if no materials or data flow between activities. The activities "receive tubes or parts" and "inform quality inspector of materials arrival" are connected by a *predecessor-successor* relationship, as they can only be carried out in sequence, the second after the first.

The process of creating workflow representations of business processes (typically called flowcharting) is "... an invaluable tool for understanding the inner workings of, and relationships between, business processes" (Harrington, 1991, p. 86). Irrespective of this opinion, however, one important point must be made about workflow representations of business processes, such as the flowchart in Figure 3.1. Although flowcharts can show the data or materials that flow between activities in a business process, these data or materials do not actually flow between activities. Hence, the data flow representation in flowcharts can be somewhat misleading.

For example, the delivery form that apparently flows between the activities "inform quality inspector of materials arrival" and "inspect materials," in reality flows between the organizational functions that carry out these activities—acquisitions assistant and quality inspector. The delivery form is a data repository that allows for the exchange of information between those two organizational functions (i.e., acquisitions assistant and quality inspector).

This shortcoming of the workflow view can be of significant importance, if the focus of a business process redesign attempt is the data flow and not the activity configuration in a business process. The workflow view "hides" information about how data flow in organizational processes (Kock & McQueen, 1996). As discussed later in this book, a focus on data flows is, in general, advisable, since the flow of data is increasingly becoming a central concern in projects targeted at boosting organizational competitiveness (Danesh et al., 2003; Kock, 2001a).

There are a number of variations of workflow representations similar to the one shown in Figure 3.1. The workflow in Figure 3.1 is an adaptation of the ANSI standard flow chart, which has been used extensively in our work with business process improvement groups (see Kock [1995] and later chapters of this book for a description of the use of this flowcharting tool in business process improvement groups). Flowchart variations are the block diagram, functional flowchart, functional time-line flowchart, and geographic flowchart (see Harrington [1991] and Harrington et al. [1998] for a more detailed discussion).

The Data Flow View

A traditional view of business processes is the data flow view, where business processes are seen as data processing entities. Data flow representations were used largely in the 1980s by computer systems analysts as an important component of what are known as structured systems analysis and design techniques (Davis, 1983; Dennis & Wixom, 2000)—a predecessor of the now relatively popular object-oriented analysis and design approach (Booch et al., 1998; Somerville, 1992).

Figure 3.2: Order fulfillment process of a central kitchen at an Italian restaurant chain (Adapted from Kock, 1995a, p. 44)

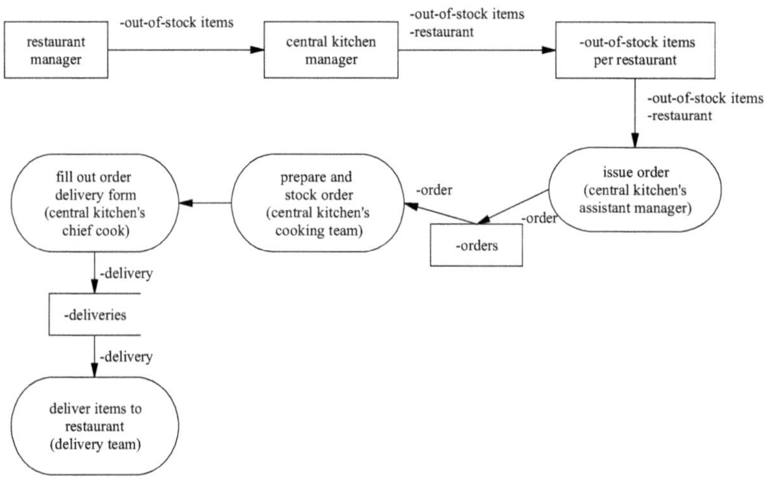

In the past, data flow representations have been used chiefly to understand the flow of data within business processes, and later to automate that flow "as is" rather than to redesign (i.e., change) business processes. This automation-of-old-processes approach was the target of strong criticism in the early 1990s, often described as the main cause of the low return on investment in information technology observed in both the 1970s and the 1980s. According to Hackett (1990), the service sector was particularly affected by this low return on investment in information technology, an assertion that seems to hold up to this day. Over the years, such return has steadily declined, even to negative figures (i.e., the investment in information technology has led to a decrease in productivity) in a number of service industries such as banking and insurance (Carr, 2003; Strassmann, 1997).

Like the workflow view of business processes, the data flow view can be expressed by a family of graphical representations, the most widely used of which is the data flow diagram, or simply DFD (Dennis & Wixom, 2000; Gore

& Stubbe, 1988; Pressman, 1987). An example of DFD obtained from the analysis of the flow of data between the restaurants and the central kitchen of an Italian restaurant chain is shown in Figure 3.2.

The shape and meaning of the symbols used in DFD charting can vary widely, depending on the business process representation standard used. In Figure 3.2, a rectangular shape represents a data source or destination—the restaurant manager and the central kitchen manager functions—in the figure. Arrows indicate the flow of data, which is described by freestanding text located beside the arrows. Oval shapes represent activities. Open-ended rectangles represent data repositories.

The business process mapped through the DFD in Figure 3.2 starts with the manager of one of the restaurants of the chain contacting the manager of the central kitchen, where all dish items are prepared. The restaurant manager tells the manager of the central kitchen that the restaurant is running short of some specific items (e.g., Bolognese sauce, spaghetti, Italian bread). The manager of the central kitchen then fills out a form on which he specifies some out-of-stock items and the restaurant that needs them, and puts this completed form into the assistant manager's inbox. Approximately every two hours, the assistant manager of the central kitchen goes through the forms in his inbox and generates and stores in his outbox the orders to be produced by the cooking team and the sequence in which they will be produced. He tries to optimize the work of the cooking team when doing this scheduling by grouping requests that require the same resources (e.g., ingredients, cooking equipment). The cooking team then collects the orders from the assistant manager's outbox and prepares the Italian dish items ordered on a first-come-first-serve basis, packing and stocking them in the delivery room as soon as they are ready. Delivery forms are filled out and attached to each of the packed items for the restaurants, which are periodically delivered by the central kitchen's delivery team.

Although incomplete models of real business processes, representations based on the data flow view of business processes such as DFDs show in a relatively clear way how data flow and are stored in business processes. As such, one can reasonably expect these representations to be more appropriate than workflow-based representations (e.g., flowcharts) in some cases. This is true especially in the analysis of business processes where the flow of data is particularly intense (Kock, 2003).

A dramatic increase in data flow has been predicted to be one of the characteristics of the present and future economies, particularly in developed and developing countries (Drucker, 1989; Toffler, 1970, 1991). Hence, it is reasonable to expect that representations based on the data flow view of business processes are more likely to be useful in business process improvement attempts than representations based on the workflow view in the business contexts of those countries. As mentioned previously in this chapter, the latter type of representations tend to provide a poorer picture of how data flow, preventing the identification of, for example, "data buffers."

Data buffers are organizational functions (rectangular shapes in DFDs) whose job in a given business process is to transfer data between other organizational functions. These "buffers" are, therefore, strong candidates to be removed from the business process and to be replaced by information technology applications. This is the case in Figure 3.2 for the function "central kitchen manager," who acts as a buffer between the restaurant manager and the assistant manager of the central kitchen. In the business process that is analyzed, the manager of the central kitchen receives data from the restaurant manager and stores it into a data repository that will be used as input by the assistant manager of the central kitchen to generate an order. A more efficient version of the business process would have the restaurant manager storing this data, with no mediation of the manager of the central kitchen, who could use his time to do other things.

When mapping business processes either through flowcharts or DFDs, one may wonder how much detail to show in the diagram. After all, the activities in a business process representation can also be seen as sub-processes themselves, which, in turn, can be broken down into new activities, and so on. In fact, seeing the activities of business processes as lower-level processes and generating more detailed diagrams by "exploding" these lower-level processes is a common practice in both flowcharting and DFD generation (Davis, 1983; Dennis & Wixom, 2000; Maull et al., 1995; Pressman, 1987). In doing so, however, two simple guidelines are suggested (Kock & McQueen, 1996).

- Each graphical representation of a business process should not have more than 15 activity (or sub-process) symbols.
- In a business process improvement context, the level of detail one should search for when modeling business processes should be defined by the breadth of improvement sought.

The first guideline is based on studies about general human cognitive limitations relating graphical representations and diagrams used in systems analysis and design (Kock, 1995a), which in turn builds on previous findings in the generic field of cognitive psychology (e.g., the "magical number seven" phenomenon discovered by Miller [1956], which led to the development of a new theory, sometimes generally referred to as cognitive chunking theory) (Glassman, 2003).

The second guideline is based on a specific concept—breadth of business process improvement (Hall et al., 1993). Roughly speaking, the breadth of improvement correlates with the number of departments affected by business process improvement decisions. The larger the breadth of improvement, the less business process detail is necessary when modeling a business process that is being targeted for redesign. If one wishes to improve business processes that cut across several (perhaps all) of the departments of an organization, the business process representation(s) used should comprise little detail about sub-processes that belong to individual departments. As a general rule of thumb, the total number of high-level business processes used to effectively represent any organizational unit is anywhere between 10 and 20 (Hammer & Champy, 1993; Maull et al., 1995).

Earlier, I mentioned that data flow representations have been used in the past chiefly to understand business processes in a pre-automation stage, so that automation of those business processes could be conducted right away and without changes made to the structure of the business processes. That approach has been traditionally favored over the approach of using data flow representations to support the redesign of business processes prior to automation.

This may be due to the orientation adopted by the designers of DFD notations and diagramming rules, some of which make it quite difficult to use DFDs in business process improvement situations. For example, representations of data flows between two rectangular shapes, as in Figure 3.2 between the restaurant manager and the central kitchen manager, violate standard DFD modeling rules. And representations of data flows between two rectangular shapes are important from the perspective of someone redesigning a business process, because those data flows are usually associated with business process inefficiencies (this point is explored in detail later in this book). Therefore, it may be necessary to adopt DFD variations that can be used more easily in business process improvement contexts. One such variation is discussed later in this book. That variation is similar to the one proposed by Kock and Murphy (2001) as part of their new methodology for redesigning specific business processes (namely acquisition processes) based on the flow of knowledge and information.

Other Business Process Views

Although the two business process views discussed—the workflow and the data flow view—are, in my opinion, the most relevant views for the purposes of this book, there are other views of business processes. Among those are the systems view and the object-oriented view.

The Systems View

The systems view of business processes is based on the traditional concept of *system*—an assembly of parts that cannot be understood only as a function of its components. A system can be defined by its emergent properties, which are system properties and, therefore, meaningless in terms of the parts that make up the system. This concept is illustrated by Checkland and Scholes.

The vehicular potential of a bicycle is an emergent property of the combined parts of a bicycle when they are assembled in a particular way to make the structured whole.

(Checkland & Scholes, 1990, p. 19)

According to the systems view, a business process can be operationally defined as an abstract entity that represents the transformation of inputs into outputs (Childe, 1995; Childe et al., 1994; Kock, 1995a). The business process's suppliers provide inputs. The business process's customers consume the outputs generated through the business process. The transformation of inputs into outputs is aimed at adding value to the customers of the business process. While there is some controversy over what inputs and outputs can be, it is reasonably safe to say that the inputs and outputs of a business process may be of three different types—goods, services, and data (Juran, 1989; Kock & Tomelin, 1996).

While philosophically appealing, the main problem with the systems view of business processes is that it adds little to our understanding of the inner workings of a business process, and, therefore, may be of little use to those who try to change the business process. According to the systems view, business processes are defined by means of sets of emergent properties that characterize them; the relationship between their components is of secondary importance.

In spite of its limitations, the systems view has proved to be more useful than the workflow view in the analysis of very complex (and often "messy") business processes, such as those related to strategic decision-making. These business processes typically cannot be analyzed as workflows, because, among other things, the number of activities and decision points required to represent them is too large to allow for effective modeling. My own experience supports this assumption. In 1992, for example, I tried to analyze one such process in an advertising company, and ended up with a very complicated model made up of more than 150 activities. One could argue that, in this case, taking a systems approach to modeling would have allowed for a better understanding of the business process. However, given time and financial constraints that are not often mentioned in popular business books, the firm's management and I simply decided to skip the strategic management process in question and move on to other more analyzable business processes.

The Object-Oriented View

One of the main proponents of the object-oriented view of business processes is Ivar Jacobson, who developed a methodology to model business processes as data objects. Jacobson's methodology was based on the concept of software object (Jacobson et al., 1995), which is a data repository with a number of operations associated to it. These operations are also called "methods" in the technical jargon of object-oriented analysis and programming. A software object typically stores data in its attributes, which are analogous to the attributes of real objects like a chair (e.g., attributes of an object "chair" would be its "color," "weight," and "number of legs" [Partridge, 1994]).

The object-oriented view is seen as an extension of the data flow view in which data repositories, often represented in DFDs by open-ended rectangles (see Figure 3.2), are permanently linked to activities that change the content of those repositories. There is a clear advantage in adopting this view. Many believe that object-oriented programming is increasingly becoming the dominant software development paradigm (it has been adopted by most of the major players in the software development industry since the 1990s). Also, the object-oriented view of business processes allows for an inexpensive transition between: (a) business process analysis and redesign, and (b) the development of new computer systems to support the implementation of the new redesigned business processes.

However, the object-oriented view has been criticized for its excessively technical orientation, preventing less sophisticated users (i.e., those who are unfamiliar with object-oriented concepts) from effectively understanding it in its full complexity and adopting it in business process improvement projects. Business process analysis and design methodologies using object-oriented representations, such as the Unified Modeling Language (UML), are still too complex to be widely accepted and used in organizations, in spite of the fact that UML has been endorsed by heavyweights of the computer community (Meyer, 1998).

This problem is compounded by the fact that less sophisticated users are often senior managers, who are usually more interested in strategic management than in technical issues, and who, therefore, do not normally have the time to become technically sophisticated enough to understand object-oriented business process analysis and design issues. The trouble with this situation is that the support of these managers is a fundamental ingredient in successful business process improvement initiatives (Davenport, 1993).

Moreover, some recent developments in the software industry have turned the building of computer systems in-house, which is facilitated by the adoption of the object-oriented view of business processes, into an often undesirable and expensive alternative. Buying off-the-shelf applications, quickly prototyping new computer applications, and outsourcing the development of computer systems to enable new organizational processes are seen by many as more desirable approaches made possible by such developments. Among these developments are the following:

- The emergence of computer-aided software engineering (CASE) tools, which support rapid application development.
- The increasing number of powerful and user-friendly general-purpose systems that can be rapidly adapted to perform a diverse range of tasks. For instance, spreadsheets can be used to build flexible small-scale customer databases and cash flow control systems, which previously would have been developed through traditional computer programming.

- The proliferation of specialized software development companies in almost all industries; companies whose software development costs are usually much lower than those of in-house development.

Summary and Concluding Remarks

In the 1990s, the business process concept became the focus of growing attention for managers, business consultants, and management researchers. This was particularly due to the business process reengineering movement, which apparently fell short of initial expectations, yet deeply influenced contemporary management thinking.

It is reasonably safe to say that, in reality, reengineering is nothing but a genre of business process improvement. Diverse business process improvement approaches may employ entirely different guidelines and analytical tools, yet all such approaches share a common focus on organizational or business processes.

A careful analysis of the history of organizational development shows that business process improvement has been an underlying force in many management movements, including the total quality management, business process reengineering, and organizational learning movements.

The central element of business process improvement is the business process, and thus it is important that we clearly define this concept. The most widely accepted definition of business process posits that it is any set of interrelated activities. However precise this definition may sound, it is broad enough to spur different interpretations or views. The two predominant business process views are the workflow and data flow views.

The workflow view focuses on the understanding of business processes as groups of interrelated activities carried out in sequence, among which data or material products are exchanged. Although it is a very intuitive and widely accepted view, it can be misleading. The fundamental problem with the workflow view is that products do not actually flow among activities. They flow among organizational functions or roles (e.g., lathe operator, inventory control manager, chief executive officer).

The data flow view focuses on how data flow within business processes, arguably without making the same mistake that the workflow view does (this view also has its limitations). Here, data are seen as moving within and outside business processes and among organizational functions. This is one of the reasons that I would recommend the data flow view over other views for business process improvement practitioners. Other reasons include its relative simplicity, its long utilization history in computer application development circles, and, finally, its support in understanding how data flow in organizations. After all, data is what mostly flows in organizations, whether they are organizations that produce or commercialize manufactured goods, information, or services.

Chapter IV

Data, Information and Knowledge

Data, Information and Knowledge are the Same Thing, Aren't They?

Quite often we hear the words data, information, and knowledge being used as if they were synonymous. But aren't data, information, and knowledge actually the same thing? And if not, what is the difference? I strongly believe that there are subtle but very important differences among these concepts, and that the nature of these differences is a relatively complex one. Moreover, from a business process improvement perspective, data, information, and knowledge serve fundamentally different purposes.

The contribution of information technology (IT) providers perhaps has been unmatched in its ability to add to our confusion over the distinction between data and information. Examples can be found in almost any specialized IT publication, conversations with IT company representatives, and even in public speeches by self-proclaimed IT gurus. For example, a senior vice-president of a large software development company was one of the keynote speakers of an international information systems conference that I attended a few years ago. He referred to the advantages of a well-known commercial group support system in the following terms:

> *... information overflow can be considerably reduced...for example, a few weeks ago I prepared a 2 megabyte report and sent it via e-mail to ten people. Each of these ten people forwarded a copy of the report to about ten other people ... as a result, my report had generated a flow of 200 megabytes of information in the network, in less than four days ...*

In this example, the speaker was referring to data, which can be measured in megabytes, as though it was synonymous with information. This can often be misleading, because a large amount of data may have very low information content, depending on how well prepared is the receiver of the data to make sense of it. Mistakenly identifying data as information is as commonplace as confusing knowledge with information.

It is curious that the confusion over information and knowledge has been nurtured by some who are widely recognized as the forerunners of the study of information and knowledge. They also have had an impact on organizations and society. One the most highly regarded personalities among management consultants and researchers, Claremont Graduate University Professor Peter Drucker (1989), describes the emergence of the information-based organization in the following terms:

> *...the business, and increasingly the government agency as well, will be* knowledge*-based, composed largely of specialists who direct and discipline their own performance through organized feedback from colleagues and customers. It will be an* information*-based organization ... Today's typical organization, in which* knowledge *tends to be concentrated in service staffs perched rather insecurely between top management and the operating people, will likely be labeled a phase, an attempt to infuse* knowledge *from the top rather than obtain* information *from below [my emphasis].*
>
> (Drucker, 1989, pp. 207-208)

If information and knowledge were the same thing, why use two words when just one would suffice? Even though information and knowledge mean different things to different people, most people use these words with different senses. The main reason these two words are often used interchangeably is because there is no agreement over their meaning.

But, who cares? Or, more precisely, why should we worry about the different nature of data, information, and knowledge? One reason is because understanding their difference may help us solve key problems that arguably result from us not differentiating among them. For example, an ocean of data may contain only a small amount of information that is of any value to us, and sifting through this ocean of data may be extremely time-consuming (Goldratt, 1991). Another reason is that, without understanding the different nature of these concepts, we cannot fully appreciate how business process improvement can contribute to knowledge sharing, a topic that I will discuss later in this book. But there are other reasons that relate to the nature of our understanding of the world, or the way we make sense of the world around us.

The world is not only what we perceive it to be through our senses; it is a combination of these perceptions and what is stored in our body, mostly in our brain, in the form of networks of neural connections (Callatay, 1986; Dozier, 1992). We can develop our neural networks by interacting with matter and living organisms, notably other human beings. However, in order to interact with other human beings, we need to externalize what is stored in our neural networks by means of a code. Other human beings should understand this code, so that communication of what is stored in the form of neural networks can take place.

If data and information are the same, how can the different information content that one e-mail message may have for different recipients be explained? More specifically, let us suppose that an e-mail message written in Spanish (a specific code) is sent to two different recipients. Let us also assume that while one of the recipients can read Spanish very well, the other cannot. In this example, the message takes up the same disk space (i.e., 3.6 kilobytes) on the computers of each of the recipients, which is a measure of the amount of data related to the message. Yet, its information content is much higher for the recipient who can read Spanish than for the recipient who cannot.

If data and information were the same, then they should not yield different amounts when measured for the same object; in this case, the e-mail message in Spanish. It is important to stress that I could have used different terms in this discussion other than data and information. For example, I could have used the terms "alpha-stractum" and "capta" instead of "data" and "information," respectively. I will stick with the more commonly used terms in this book—data and information—because I believe that the sense in which I have just used these two terms is their most usual sense.

The distinction between knowledge and information is a bit more abstract than between information and data. In order to make this distinction as clear as possible, let us consider the following dialogue between a doctor (D) and her patient (P).

D: *So, what brings you here today?*

P: *I don't know doctor. I've been feeling a bit strange in the last couple of weeks.*

D: *What do you mean by "strange"?*

P: *Burning eyes, stuffed nose...and these things go and come several times a day.*

D: *Any headaches or fever?*

P: *No, not at all.*

D: *Well, we'll run a checkup on you, but I think you probably have an allergy.*

The patient was feeling the symptoms of what could be an allergy, and, therefore, he went to see his doctor, an expert who likely knows more about medicine than the patient. The patient described his symptoms, and the doctor made a tentative diagnosis—you probably have an allergy.

Is what the patient told the doctor enough for someone without medical expertise to make the same tentative diagnosis? If this were the case, very few people would agree to pay doctors for consultations. Doctors possess more of something that patients do not have, something typically referred to as knowledge, in the specific field of medicine.

Is the nature of the expert knowledge possessed by the doctor, in this case, the same as that of the perception of symptoms experienced by the patient? No, for the simple reason that expert knowledge can be used to generate conclusions based on the description of symptoms, something that the descriptions alone cannot. Therefore, the natures of descriptions and expert knowledge are different, and it can be shown that none of them is the same as data. This also suggests that the descriptions are instances of something unique, which I refer to here as information.

Data are Carriers

The usual sense of the term *data*, even if not explicitly stated, is that of carriers of information and knowledge. Data flow in organizational processes between the functions that carry out process activities. This flow takes places through various media, particularly paper, digital electrical impulses (e.g., electronic data interchange systems), analog electrical waves (e.g., telephone), electromagnetic waves (e.g., radio), and air vibrations (e.g., face-to-face conversation). Data can also be stored for later use in different storage media such as magnetic media (e.g., hard and floppy disks), paper, and volatile digital memories (e.g., random access memory, or RAM, in personal computers).

Data either are transferred or stored through a process of changing or generating perturbations on a given communication or storage medium. A blank sheet of paper, for example, can be used for data storage (i.e., to write down an address of a friend) or for transfer (i.e., to write a memo to an employee) by applying ink on it. From a more business-oriented perspective, if a machine operator wants to tell the supervisor about a problem with a metal-shaping machine, the operator can approach the supervisor and speak face-to-face. In doing so, the operator uses vocal cords to generate vibrations in the air (volatile data) that will be received and decoded by the recipient through hearing organs.

The main point that I want to make here is that data will only become information or knowledge when it is interpreted by human beings (Kryt, 1997) or, in some cases, artificial intelligent agents (Russel & Norvig, 1995). As data can be stored and transferred by business process functions through applying changes to storage and communication media that will be interpreted by other business process functions, we can try an operational definition within the context of business process management.

Let us assume that John performs an organizational function, i.e., he carries out an activity in an organizational process. We can say then that data are permanent or volatile changes applied to a communication medium by John to store or transfer information or knowledge. John, someone else, or perhaps an artificial intelligent agent, will later use these to perform an organizational activity.

The measurement of data depends on the medium used to store or transfer it, as well as on the code used. In most organizational processes, data can be measured in words or symbols when the medium used is paper, and bits or bytes (one byte is a group of eight bits) when the medium used is a digital one.

In many ways, a bit can be considered the smallest and most fundamental unit of data. It can take only two values: 0 (false) and 1 (true). A group of eight bits forms a byte. And, since the number of possible bytes is 2^8 or 256, there can be a direct correspondence between bytes and certain symbols (e.g., the letters of the English and other alphabets). One such set of symbols, largely used to convert alphanumeric characters into bytes and vice-versa, is called the American Standard Code for Information Exchange (ASCII) code. Most operating systems in personal computers use the ASCII code, or an extended version of it, to map symbols that have meaning to human beings (e.g., letters and numbers) into bytes stored in any of the computer's data storage devices (i.e., RAM, hard disk, etc.).

Information is Descriptive

A hot issue in business circles since the 1990s has been the advent of the "information society," the "information era," and the "information-intensive" organizations. However, any discussion regarding these issues should, of necessity, focus on the nature of information. What is it? Is it a specific kind of entity? If yes, how can we differentiate information from other similar entities? These are core questions in the continuing debate within a number of disciplines such as information systems, management science, engineering, and philosophy. A substantial portion of the literature in these disciplines is devoted to defining information. However, as Budd and Raber note:

> *In the course of doing so [i.e., defining information], many aspects of information (technical, physical, semantic, epistemological) are featured as part of the discussion. Part of what emerges is a multifaceted idea and thing that is, at times, defined in terms of what it is not. For instance, information is not merely data; organization and intended meaning transform the bits of data into something that can inform.*
>
> (Budd & Raber, 1996, p. 217)

From a business process-oriented view, information can be seen as carried by data, and as being eminently descriptive. From a linguistic perspective, the typical instance of information is the utterance called assertion. One example of assertion is, "Today is a sunny day." Independently of what this assertion means exactly (the word sunny may mean different things to different people, from sparsely clouded to clear-sky weather), it provides a description of the current state of the environment surrounding us. If the environment is seen as an object, the assertion can be seen as defining an attribute of the object—in this case, the weather—as sunny.

Information can be qualified in different ways; it can be more or less complete or accurate, and it can refer to the past, present, and future. For example, the assertion, "Today is hot!" conveys less accurate information than the assertion, "Today's temperature is 85 degrees Fahrenheit." Both assertions describe the present—today. The assertion, "The temperature on this day during the last three years has averaged 87 degrees Fahrenheit," provides information about the past. The assertion, "Tomorrow the top temperatures will be in the low 90s," provides a description of the future. Although similar to descriptions of the past and the present, descriptions of the future, by their own nature, always carry a certain degree of uncertainty.

Knowledge, which will be discussed in more detail in the next section, is often used to generate more information, based on information at hand. The information thereby generated (or inferred) is usually not obvious, and, therefore, possesses some added value in relation to the primary information received as an input by the knowledge holder. One example is the generation of information about the future i.e., the weather in New York tomorrow) based on information about the present and past (i.e., the weather patterns in New York during the last two years) up to now. This type of information about the future is produced by meteorologists, based on their knowledge about the science of weather forecasting. It is then bought by news services, which in turn broadcast that information to their audiences and, in the process of doing so, manage to make some money.

The Value of Information

One interesting aspect of information is that its value (i.e.,how much someone is willing to pay for it and can benefit from it) in general seems to directly correlate to some of its attributes. Among those attributes are the following:

- Its advanceness—how much time in advance it describes the future
- Its accuracy—how accurate the description is
- Its completeness—how complete the description is

Let me explain the different nature of these attributes in a business context. The corporate war between Coca-Cola and Pepsi in the 1980s was largely one of product differentiation (Ramsey, 1987). Both Coca-Cola and Pepsi tried to increase their shares of the "cola" soft drink market by launching new differentiated (e.g., diet) products ahead of each other. Consider the similar situation of two companies—A and B—competing for two million customers in the same industry. Each customer consumes a product supplied by both companies. Analogous to the cola war, the product is essentially the same, but the main difference is the brand. Each customer consumes 70 units of the product (at $3 each) every year, making it a $420 million per year market. Company A has 90% of the market, $378 million, while Company B has the other 10%, or $42 million. Both companies sell with a pre-tax profit margin of 17%, which yields approximately $64 million for Company A and $7 million for Company B in absolute pre-tax profits.

Now suppose that Company B decides to launch a new product in the market, whose development time is approximately nine months. The product has the potential to bring Company B's market share up to 20% and send Company A's share down to 80%. This would raise Company B's absolute pre-tax profits up to about $14 million and make Company A's profits plummet to nearly $57 million. From Company A's perspective (and the value of information always depends on its users and their context), one piece of information can make a lot of difference—the information that Company B is going to launch a new product.

This piece of information can have a high advanceness, if it is provided to Company A well in advance of the product launch, enabling it to take appropriate countermeasures. The same piece of information can have a high accuracy, providing accurate details about the product that is going to be launched (e.g., it might include the precise date of launch). The information can also have high completeness, providing a rich description of the new aspects of the product (e.g., the new flavor, amount of saturated fat, sweetener used, etc.).

If Company A has no access to information about the new product launch and, for instance, obtains some imprecise information a few weeks before the new product is launched, it will have to endure a loss in pre-tax profits of $7 million; this is the worst-case scenario. However, if it gets its hands on accurate and complete information early enough, it can take preventive measures to at least reduce its losses. For example, if the information is obtained more than nine months in advance (i.e., has high advanceness), but leaves uncertainty about the characteristics of the product (i.e., has low accuracy and completeness), then Company A might have to develop a range of new products to dampen Company B's new product's potential impact on market share. Its profits still may be reduced due to increased product development costs, but not as much as in the scenario in which no information about the competing company's launch was available.

Having access to detailed information about Company B's new product (i.e., highly accurate and complete information) only four months before the launch (i.e., low advanceness information) may lead to a similar end result. That is, Company A may be able to develop an intermediary product that will reduce Company B's new launch's impact on market share.

Figure 4.1. The value of information

	Low accuracy and completeness	**High accuracy and completeness**
High advanceness	Medium value	High value
Low advanceness	Low value	Medium value

The best scenario is perhaps one in which Company A has access to highly accurate and complete information about Company B's new launch early enough (i.e., the information has high advanceness) so it can develop a similar new product and get it out into the market before Company B does. According to our initial assumptions, this could potentially bring Company A's market share up to 95% and increase profits by about $4 million.

In this example, no information or information with low accuracy, completeness, or advanceness would be of low value to Company A. Information with high accuracy and completeness, but low advanceness (or vice versa) would have a medium value, as it could prevent a loss of $7 million in pre-tax profits a year. Finally, information with high accuracy, completeness, and advanceness would have a high value, enabling an increase in profits of $4 million a year. This relationship between information value and its attributes is illustrated in Figure 4.1.

Although the example is concerned with a decision-making process at the strategic level, we can extrapolate the relationship among information value and the attributes advanceness, accuracy, and completeness to most organizational processes. Simply put, business process-related information seems to be an important enabling factor for the members of a business process team (i.e., those who perform business process activities) to do their job efficiently and effectively, whatever the business process is.

Knowledge is Associative

While information is eminently descriptive and can refer to the past, present, and future, knowledge is, by its own nature, eminently associative. That is, it allows us to associate different world states and respective mental representations, which are typically linked to or described by means of pieces of information. In other words, knowledge allows us to link different pieces of information and make decisions based on those linkages.

The associative aspect of knowledge is of two main types—correlational and causal—which are, in turn, only two of the types referred to by Weick and Bougon (1986) as cognitive archetypes. And again, human beings can directly store knowledge through neural connections, which in turn are concentrated mostly in the brain. If someone loses part of his or her brain, that person may also lose part of the knowledge previously stored there in the form of neural connections.

Correlational knowledge usually connects two or more pieces of information that describe events or situations that have happened, are happening, or will happen at the same time. Causal knowledge connects pieces of information that describe the state of the world at different times. For example, consider the associative knowledge represented in the following decision rule: "If John has a fever and is sneezing, then John likely has a cold." The knowledge embodied in this decision rule is of the correlational type, because it affirms that someone who has fever and is sneezing is, in fact, displaying typical cold symptoms. That is, having fever, sneezing, and having a cold typically happen at the same time.

Another example of a different type of knowledge is provided by this rule: "If John smokes a pack of cigarettes a day, then he will probably die from lung cancer within 20 years." This decision rule expresses causal knowledge. As such, the rule connects two events that take place at different times: John smoking a lot in the present, and John dying of lung cancer in the future. It is to causal knowledge that Dennett (1991) refers, when he points out the following:

The brain's task is to guide the body it controls through a world of shifting conditions and sudden surprises, so it must gather information from that world and use it swiftly *to "produce future" – to extract anticipations in order to stay one step ahead of disaster [original emphasis].*

Knowledge drives the flow of myriad decisions that have to be made, even in the simplest organizational processes. Steel plants, for example, rely on business process teams to load and operate smelters. Consider the predictive knowledge expressed in the rule: "If the smelter is set at a temperature of 3,000 degrees Celsius, then a one-ton load of steel will be smelted in 43 minutes." This is one of the pieces of knowledge that allows a smelter operator to predict that a batch of solid steel weighing about one ton will be in liquid form approximately 43 minutes after it is loaded into the smelter, if the smelter is set properly. This prediction allows the smelter operator to program a stop in the smelting process at the right time and let the liquid steel flow out of the smelter, which saves energy and, at the same time, prevents the steel from overcooking.

In order for teamwork to yield effective and efficient outcomes, those who perform activities in a business process must share predictive knowledge. In the example, those who use the steel in liquid form for shaping steel parts should ideally hold at least part of the knowledge held by the smelter operator. If they know about the 43-minute rule, they can also predict that a batch of steel will be ready within 43 minutes from the time it is loaded in solid form, and they can have their own equipment prepared at the right time to work on the liquid steel.

In business in general, knowledge seems to be inextricably linked to decision-making (Holsapple & Whinston, 1996; Kock & Davison, 2003; Olson & Courtney, 1992), perhaps because one of the best ways to assess the actual value of knowledge is through the assessment of the outcomes of decisions made based on it. Holsapple and Whinston (1996) discuss the importance of knowledge for decision-making.

> *For centuries, managers have used the* knowledge *available to them to make* decisions *shaping the world in which they lived. The impacts of managers' decisions have ranged from those affecting the world in some small or fleeting way to those of global and lasting proportions. Over the centuries, the number of decisions being made per time period has tended to increase. The complexity of decision activities has grown. The amount of knowledge used in making decisions has exploded. There is no sign that these trends are about to stop. If anything, they appear to be accelerating [original emphasis].*

Knowledge has been distinguished from information and also linked with decision-making in different fields of research and academic disciplines. In the field of artificial intelligence, for example, information typically has been represented by what has been referred to as "facts," which are essentially assertive statements that describe something. Knowledge, on the other hand, has been expressed by means of a number of different representations, such as semantic networks, frames, scripts, neural networks, and production rules; the latter being the most common in practical knowledge-based computer systems (Callatay, 1986; Holyoak, 1991; Olson & Courtney, 1992). Production rules are conditional statements in if-then form, like the ones used to exemplify knowledge in this section.

In the fields of psychology and social cognition, knowledge has been expressed through schemas (Lord & Foti, 1986) and cognitive maps (Weick & Bougon, 1986). These, in turn, are seen as guiding individual and group behavior, and used in the processing of environmental stimuli-related inputs obtained through the senses.

The concept of schema was developed as a reaction to studies of memory pioneered by Ebbingaus, which made use of arbitrary materials and sensorial stimuli to determine factors that influence the formation of memory and recall of information (Gardner, 1985). The development of the concept of schema is credited to Bartlett (1932), who used a North American Indian folk tale called "The War of the Ghosts" to show that existing mental structures strongly influenced memory formation and recall. Such existing mental structures, which were used by Bartlett's study subjects to process information coming from the tale, were called schemas. Essentially, Bartlett has shown that individuals possessing different schemas would interpret the tale, which is filled with strange gaps and bizarre causal sequences, in substantially different ways.

In biology in general, and, more particularly, in neurology, knowledge is typically seen as associated with long-term nerve-based memory structures whose main goal is information processing (Pinker, 1997). Information is seen as being usually associated with short-term neural connections that appear to vanish from conscious memory after a while. For example, the knowledge of how to operate a telephone is stored in long-term memory structures, whereas the information represented by a phone number is stored in short-term memory structures.

The Value of Knowledge

Knowledge is usually much more expensive to produce than information. For example, information in the form of mutual fund indicators (e.g., weekly earnings, monthly price fluctuation) is produced by means of little more than simple calculations performed on data about share prices and their fluctuation over a time period. The knowledge of how mutual fund indicators fluctuate, however, requires years of analysis of information, which leads to the development of knowledge that allows an expert investor to select the best mutual funds in which to invest money, given a particular configuration of the economy. Which leads us to the question: How is knowledge produced?

Comparative studies of experts and non-experts suggest that expertise usually is acquired through an inductive process in which generalizations are made based on the frequency with which certain pieces of information occur in a combined way. These generalizations are the basis for the construction of knowledge (Camerer & Johnson, 1991).

A different and less common method used to generate knowledge is deduction, whereby hidden knowledge is produced based on existing knowledge through a set of logical steps (Teichman & Evans, 1995). This method has been used in the development of a large body of knowledge in the form of theorems, particularly in the fields of mathematics and theoretical physics (Hawking, 1988).

An example of knowledge building through induction is that undertaken by novice investors in the stock market. The observation that shares of a small number of companies in high technology industries have risen 10 percentage points above the Standard & Poor's 500 average index during a period of six months, may prompt a novice investor to put all monies into these shares. A professional investor, however, knows, based on probably 10 years of experience as a broker in the stock market, that a six-month observation period is not long enough to support such a risky decision, and opts for a more diversified portfolio. In cases such as these, the novice investor will eventually lose money, particularly because the decision to sell will probably follow the same pattern as the decision to buy. It will be based on inferences that are based on a time span that is too short, leading an investor to buy shares that are overvalued and sell these shares when they are undervalued. According to Boroson (1997), who studied investments strategies extensively, most non-professional investors follow this recipe, which, in most cases, leads to negative and, in some cases, disastrous consequences.

This example illustrates a key finding from research on cognitive psychology—people usually tend to infer knowledge based on the observation of a small number of events; that is, on limited information (Feldman, 1986). Moreover, once knowledge structures are developed, changing these structures can become more difficult than developing them from scratch (Woofford, 1994).

A conversation that I had some time ago with a university colleague illustrates some of the cognitive biases already discussed. My colleague had gone to two different agencies of the New Jersey Motor Vehicle Services, where he met employees who lacked sympathy and friendliness. He also had gone to a similar agency in the state of Pennsylvania, whose employees he found to be very nice. Later, during a chat with friends, he said:

All MVS [i.e., Motor Vehicle Services] employees in New Jersey are very grumpy; difficulty to deal with ... The state of Pennsylvania is much better in that respect...

I pointed out that he had just made a gross generalization, given the small sample of MVS agencies visited—two in New Jersey and one in Pennsylvania. Although he agreed with me, he was, nevertheless, adamant that he would never go to a New Jersey MVS agency again, unless it was absolutely necessary. If this was the case, he said he would ask a less touchy person to go—his wife.

The development of theories of knowledge (also known as epistemologies) and scientific methods of inquiry has been motivated by a need to overcome cognitive biases such as those already illustrated. This has been one of the main common goals of thinkers such as Aristotle, René Descartes, Gottlob Frege, Bertrand Russell, Karl Popper, and Thomas Kuhn. Epistemologies and scientific methods have provided a basis for the conduct of research in general, and in consequence for technological advances that have shaped organizations and society. Every year, hundreds of billions of dollars are invested in research, with the ultimate goal of generating highly reliable and valid knowledge. And the market value of organizations is increasingly assessed, based on the amount of knowledge that they possess, rather than on their material assets base (Davidow & Malone, 1992; Kock & Davison, 2003; Toffler, 1991).

Paul Strassmann (1997), a former information technology executive at companies such as Xerox, Kraft Foods, and the US Department of Defense, suggests that variations in the perceptions of organizational knowledge account for the growing trend toward overvaluing, (or undervaluing) stocks of publicly traded companies. According to Strassmann, the perception that a stock is overvalued stems from the failure of current accounting systems to account for the knowledge assets of organizations, and he presents an impressive array of data to support this idea. Abbott Laboratories is one of the companies he used to illustrate this point.

Over a period of seven years from 1987 to 1994, the ratio between Abbott's market value (defined by stock price) and its equity swung from five up to nearly eight, and back down to about seven. However, the ratio between market value and equity plus knowledge assets (which correlates the knowledge capital of a company) remained almost constant over that period, smoothly gravitating around two. This supports Strassmann's position that the market perceives the accumulation of knowledge assets, which is reflected in the high correlation between share prices of organizations and their knowledge assets, even though the knowledge assets are not shown on a company's balance sheet.

The sustained stability of the market-to-capital ratio which accounts for the steady rise in the knowledge capital of Abbott Laboratories confirms that the stock market will recognize the accumulation of knowledge as an asset even though the accountants do not. The stock market will also reward the accumulators of knowledge capital because investors recognize that the worth of a corporation is largely in its management, not its physical or financial assets.

(Strassmann, 1997, p. 13)

When we move from a macroeconomic to a microeconomic perspective and look at the business processes of a firm, the trend toward valuing knowledge seems to be similar to the one just described. Causal knowledge allows for the prediction of business process-related outcomes, ranging from more general predictions (i.e., prediction in connection with a group of customers' acceptance of a new product) to much more specific predictions (i.e., the need for slight manual corrections on a computer board surface after it goes through an automatic drill). Correlational knowledge enables process-control workstation operators at a chemical plant to link a sudden rise of an acidity gauge to an incorrect setting of the flow through a pipe valve. This enables the operators to take the appropriate measures to bring the acidity level down to normal.

The workers who hold bodies of expert knowledge are rewarded according to their ability to use them to perform business process activities in an efficient (e.g., low-cost) and effective (e.g., high-quality) way. This is typically done through linking different types of information, which can be done through formal education or personal experience (i.e., the building of mental knowledge bases), and generating information about the future based on information about the past or present (i.e., predicting the future).

Organizational wealth is closely linked to the ability to build and use technological artifacts to control future states of the economic, physical, and other environments in which organizations operate. However, this control is impossible without the related ability to predict the future, which, in turn, relies heavily on predictive, or causal, knowledge.

Organizational knowledge is believed to be the single most important factor that ultimately defines the ability of a company to survive and thrive in a competitive environment (Davidow & Malone, 1992; Drucker, 1995; Kock & Davison, 2003; Kock & Murphy, 2001). As previously mentioned, that knowledge is probably stored mostly in the brains of the workers of an organization, although it may also be stored in computer systems and databases (Alster, 1997; Strassmann, 1996, 1997), and other archival records (e.g., printed reports).

Whatever form it takes, knowledge is a commodity. And, as such, it can be bought and sold, which makes its value fluctuate more or less according to the laws that regulate supply and demand. Abundant knowledge, which can be represented by a large number of available professionals with the same type of expertise, becomes cheap when supply surpasses demand, which is typically reflected in a decrease in the salaries of some groups of professionals. On the other hand, a situation in which some types of highly specialized knowledge are in short supply, while demand grows sharply in a short period of time, can lead the knowledge holders to be caught by surprise when faced with unusually high bids from employers. For example, some Web Java programmers were being offered yearly salaries of up to $170,000 early in 1996, even though the demand for their new expertise was virtually nil until 1995. This was the year Java was first released by Sun Microsystems; two years after the University of Illinois began the distribution of its Web browser Mosaic. Around the end of 2002, a Web Java programmer would be lucky to get a job offer at all.

Linking Data, Information and Knowledge

Although they are different conceptual entities, data, information, and knowledge are inextricably connected. This may be one of the reasons they are so often confused. As discussed before, data are perturbations on a communication or storage medium that are used to transfer or store information and knowledge. Therefore, knowledge and information can be neither communicated nor stored without data.

Information is used to describe the world, and can provide a description of the past, present, and future. Unlike information about the past and present, information about the future always carries a certain degree of uncertainty.

Correlational knowledge allows for the linking of different pieces of information about the present. Usually, some of the information pieces are obvious and used as a departure point, while the other pieces are hidden and allow for relevant decisions. Causal, or predictive, knowledge, on the other hand, enables the production of information about the future, typically based on information about the past and/or the present.

That is, information is generated based on both correlational and predictive knowledge. However, the reverse relationship is also valid; that is, knowledge can be generated based on information. In fact, the main means by which reliable knowledge is produced is the systematic analysis of information about the past and present. This analysis typically leads to the observation of patterns that are combined into predictive and associative rules (i.e., knowledge).

Consider, for example, the following case involving Hopper Specialty and NCR (Geyelin, 1994). In 1987, Hopper Specialty, a retail vendor of industrial hardware in Farmington, New Mexico, decided to purchase a computerized inventory management system from NCR, a large developer of computer hardware and software headquartered in Dayton, Ohio. The system, called Warehouse Manager, was installed in 1988. Several problems surfaced immediately after the system was installed.

According to Hopper Specialty's representatives, the system never worked as it was supposed to, displaying an assortment of problems such as extremely low response times, constant locking up of terminals, and corrupted data files. In 1993, more than five years after the system was installed, Hopper Specialty cancelled the contract with NCR and sued the company, claiming that it had suffered a loss of $4.2 million in profits due to problems caused by the installation and use of Warehouse Manager. NCR's lawyers immediately and successfully filed a motion requesting that the lawsuit be dismissed on the grounds that it was filed too late—New Mexico's statute of limitations for this type of lawsuit is only four years.

Ethical considerations aside, NCR's lawyers had access to information and knowledge that allowed them to safely move for a case dismissal. This information was New Mexico's statute of limitations, which can be expressed by the following assertion: "In New Mexico, a lawsuit such as the one filed by Hopper Specialty should be filed within at least four years after the alleged breach of contract occurs." The knowledge possessed by NCR's lawyers allowed them to build a link between information about the law (in this case, the

Figure 4.2. The relationship between data, information and knowledge

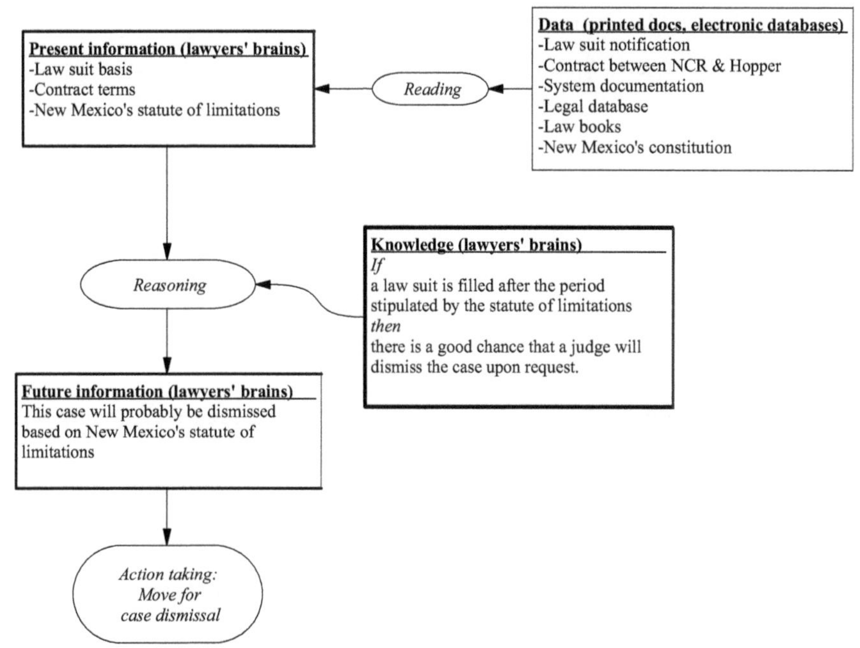

statute of limitations) and the likely consequence (information about the future) of grounding their defense on New Mexico's statute of limitations. This knowledge can be summarily expressed by the rule, "If we move for a case dismissal based on New Mexico's statute of limitations, then it is likely that the case will be quickly dismissed by the judge presiding the case."

Figure 4.2 depicts the relationship among data, information, and knowledge, based on the case discussed. The following printed or electronic documents store information that could be used by NCR's lawyers to defend their company in the lawsuit filed by Hopper.

- The lawsuit notification
- The contract between NCR and Hopper
- Warehouse Manager's documentation
- A legal database of previous cases
- Law books
- New Mexico's constitution

The items in Figure 4.2 are physical or electronic records (i.e., data), which first had to be read by NCR's lawyers so they could extract some pieces of relevant present information (i.e., information about the present situation). Examples of such pieces of relevant information are the terms of the contract between NCR and Hopper, and New Mexico's statute of limitations.

Present information can then be combined with knowledge linking the main goal of a generic statute of limitations and the likely consequences of anyone not observing the lawsuit filing expiry period stipulated by it. This combination of knowledge and information allows for the prediction of the future with a certain degree of certainty; that is, the generation of at least one likely future scenario, or information about the future. In the case of NCR versus Hopper, this future information was the prediction that the presiding judge would dismiss the case based on New Mexico's statute of limitations. NCR's lawyers, therefore, took the appropriate action of moving for a case dismissal.

Summary and Concluding Remarks

There is generalized confusion about the concepts of data, information, and knowledge, which is largely fueled by disparate meanings assigned to these concepts by diverse academic disciplines and industry groups. Although many use the three words synonymously, there are subtle and relevant differences among data, information, and knowledge.

Data are carriers of information and knowledge. Information and knowledge are coded into data so they can be either synchronously communicated (i.e., communicated in a situation where both sender and receiver interact at the same time) or stored for future use. The future use can be accomplished by the same person who generated the data, or by other people. In the latter case, data is used for asynchronous communication, which is defined as communication that is time-disconnected.

In organizations, data flow between functions that carry out business process activities. Such flow can take place through various media, particularly paper, digital electrical impulses (e.g., electronic data interchange systems), analog electrical waves (e.g., telephone), electromagnetic waves (e.g., radio), and air vibrations (e.g., face-to-face conversation). Data can also be stored for later use in different storage media such as magnetic media (e.g., hard and floppy disks), paper, and volatile digital memories (e.g., RAM memory in personal computers).

Information is predominantly descriptive and refers to the past, present, and future. While information about the past and present can be completely accurate, information about the future is always inherently uncertain. The most typical instance of information, from a syntactic perspective, is the linguistic construct known as assertion, which has the general form—object-verb-attribute. One example of assertion is "Today is a sunny day," where the object is "today's weather," the verb is "is," and the attribute is "sunny."

Unlike information, knowledge is chiefly associative. That is, it defines associations between pieces of generic or specific information so that hidden information instances can be inferred from information at hand. Knowledge associations can be correlational or causal. Correlational knowledge associations link two or more pieces of information related to same-time events. Causal knowledge links information instances that relate to different-time events.

The economic value of both information and knowledge can be estimated based on the decisions enabled by them. Information's value varies according to a number of attributes, including its accuracy, completeness, and advanceness (i.e., how much time it is obtained before the event to which it refers). The economic value of knowledge is more difficult to estimate. And, given the fact that knowledge is harder to acquire than information, it is reasonable to speculate that, when compared with information under the same general conditions, its economic value usually will be higher.

Evidence suggests that people who perform specialized functions in the same organizational process should share a certain percentage of their individual knowledge. Although many formidable obstacles to business process-related knowledge sharing exist, it is an important ingredient in productivity and quality optimization. This is an issue that will be discussed in more detail later in this book.

Chapter V

Business Process Improvement and Knowledge Sharing

Organizational Knowledge and Competitiveness

As we have seen earlier in this book, knowledge, whether stored in the brain, computer databases, or other storage media, is more often than not used for the processing of information. Information processing, in turn, has been identified as the main reason organizations exist[1] (Galbraith, 1973). That is, purposeful organization of people, capital, and other resources is necessary so information processing can be done efficiently and effectively. Information processing, in turn, is seen as a fundamental step in the generation and delivery of products and services by organizations to their customers.

Given the prominent role that information processing seems to play in organizational processes, and the assumption that information processing relies heavily on knowledge, the frequent claims that the collective knowledge held by organizations is the single most important factor defining their competitiveness do not seem unreasonable. The amount of relevant shared knowledge among individuals in business process teams has been linked to the efficiency and effectiveness of such teams (Boland & Tenkasi, 1995; Nelson & Cooprider, 1996; Nosek & McNeese, 1997). Shared team knowledge has been equated to higher flexibility and efficiency of organizational processes, as it can reduce the need for bureaucratic and automated procedures to mechanize and standardize procedures (Davidow & Malone, 1992). That is, more shared knowledge among team members may reduce the need for workflow control and automation, making business processes more efficient.

But what is organizational knowledge, and how is it related to team knowledge? Knowledge exists in organizations in a dispersed way, and is predominantly held by the individuals who perform business process activities. A concept that tries to expand the locus of knowledge, from the individual towards the group, is the concept of team knowledge (Katzenbach & Smith, 1993). Team knowledge is defined as the collective knowledge possessed by groups of individuals involved in the execution of organizational processes, regardless of business process scope. Such business processes can be as diverse as the processes of home loan approval and hamburger preparation.

An even higher level concept has been created to refer to the collective knowledge of an organization (i.e., organizational knowledge or "knowledge of the firm") (Davenport & Prusak, 2000; Kogut & Zander, 1992), which can be defined as the combined knowledge of the various business process teams that make up an organization. Part of this collective knowledge also can be stored in data storage devices, often as components of computer-based systems (Kock & Davison, 2003; Strassman, 1996).

The Need for Knowledge Sharing

Due to its associative nature, the continuous buildup and intensive use of knowledge is a necessity in a complex society. Here, the term complexity implies a large number of associations or interdependencies, whether we look at society from an environmental, artifact-oriented, sociological, psychological, or any other relevant perspective (Gleick, 1993; Lewin, 1993; Stacey, 1995).

Knowledge creation feeds complexity and vice-versa (Probst & Buchel, 1997) in what could be seen as an open-ended spiral. For example, new discoveries about a terminal disease and its genetic roots can trigger the development of new technologies and drugs for treatment and prevention of the disease in question. This, in turn, can lead to the development of new equipment and, on a different scale, new drug manufacturing companies. New governmental market regulations may follow. New militant groups fighting for their rights may emerge as those who have the genes that cause the disease organize themselves against possible discrimination by insurance companies. New research fields, theories, and academic disciplines may be spawned. Eventually, the initial discoveries about the terminal disease and its genetic roots could lead to the development of knowledge and complexity at many levels, some of which were entirely unforeseen by the those who made the original discoveries.

As knowledge becomes more voluminous and complex, so does the need for knowledge specialization by individuals. Through formal and informal education, as well as practice, experts in fields as diverse as accounting and medicine absorb and use specialized knowledge that is not held by large sections of the population in general. The market rewards knowledge specialization and task-relevant expertise through higher paying jobs and social status, in a way that generally follows the law of supply and demand (Huttunen et al., 2001). That is, specialized knowledge that is in high demand and low supply tends to be highly compensated.

Obstacles, particularly in the form of time constraints, prevent individuals from becoming experts in several different knowledge specialties at the same time, and from acquiring expertise in certain tasks too fast. For example, earlier studies by Simon and Chase (1973) suggest that a chess player cannot reach the grandmaster level in fewer than nine or 10 years, regardless of how hard he or she tries and how intelligent he or she is. This is a very interesting finding that has puzzled many researchers over the years, but that has been empirically demonstrated many times over. In fact, many experts who mentor others know the likely amount of time that their students will need to acquire a certain level of expertise in a particular task, whether the task is related to a particular form of artistic expression, sport, or work specialty.

A large and highly educated mass of people spanning many countries ensures that knowledge is created at a very fast rate so as to push individuals into focusing their cognitive efforts on narrow fields of expertise. The more educated people there are, the higher is the rate at which knowledge is created, and the stronger is the pressure toward specialization. The growth in the amount of knowledge related to previously well-defined disciplines forces those disciplines to be broken down into subdisciplines with their own bodies of knowledge.

As knowledge becomes more specialized, so does the need for information and knowledge sharing, which can be achieved through oral and written communication among those who possess different pieces of specialized knowledge. This need is motivated by the fact that even though knowledge tends to grow very specialized (or precisely because of that), most processes in society require the engagement of several individuals, each of them contributing their own piece of knowledge. In organizations, as well as in society in general, knowledge to carry out processes is often not found in concentrated form. Hayek points this out in his seminal article, "The Use of Knowledge in Society."

The peculiar character of the problem of a rational economic order is determined precisely by the fact that the knowledge of the circumstances which we must use never exists in concentrated or integrated form but solely as the dispersed bits of incomplete and frequently contradictory knowledge which separate individuals possess.

(Hayek, 1996, p. 7)

An analysis of the shop floor of two automobile manufacturers provides a good illustration of the distributed nature of knowledge. Volkswagen and Ford's plants in Sao Paulo manufactured several car models in the 1990s. Although Volkswagen and Ford usually designed each of the models assembled in their plants, most of the parts that went into the models came from their suppliers, which could easily amount to several hundred for each automobile manufacturer. Breaks, engine parts, exhaust pipes, seat belts, and the like were individually obtained from different suppliers and assembled into a car by the automakers.

Among the reasons that outsourcing the manufacturing of car parts was (and still is) more economical for Volkswagen and Ford than making those parts in-house, is that the cost of keeping and managing the specialized knowledge that went into economically and effectively building each car part was too high.

Outsourcing pushes the responsibility of keeping and managing part-specific knowledge to the suppliers. But, although each supplier possesses the knowledge that goes into manufacturing each car part, Volkswagen and Ford's engineers need to hold part of that knowledge in order to design their cars. For example, they need to know whether an airbag, which is manufactured elsewhere, will inflate according to certain specifications if they reduce the size of the airbag's compartment. That is, sharing knowledge becomes a necessity if the automobile manufacturers and their suppliers are to build low-cost cars that meet car buyers' expectations (Kock, 1995a; Parente, 2003). And such expectations are likely to be increasingly inflated in a highly competitive industry, such as the automobile manufacturing industry.

Organizational Learning and Knowledge Transfer

The example of the automobile manufacturers highlights the need for knowledge sharing among different organizations—the manufacturers themselves and their suppliers, in this case. One management movement has consistently argued for the development of knowledge creation and sharing capabilities within and among organizations as a fundamental step towards achieving heightened competitiveness. This management movement is the organizational learning movement (Garratt, 1994; Kofman & Senge, 1993; Senge et al., 1994).

In order to foster knowledge creation and sharing, it is argued that learning organizations should establish an organizational culture that is conducive to those activities that promote knowledge creation and sharing. A climate of risk-taking and experimentation has been found to be an important factor in establishing such organizational culture (Senge, 1990). That climate can be achieved through the adoption of new management practices and paradigms that stimulate creativity and proactive behavior (Nevis et al., 1995), as well as social interaction (Roskelly, 1994).

In spite of attempts to create organizational cultures conducive to learning, the transfer of acquired knowledge from one part of an organization to another remains a complex and problematic issue in most organizations (CHE, 1995; Davenport & Prusak, 2000; Kock & Murphy, 2001), a situation that hasn't shown many signs of improvement over time. This is particularly unfortunate, as the transfer of acquired knowledge across different organizational areas has been presented as one of the most important components of organizational learning (Redding & Catalanello, 1994) and competitiveness (Boland & Tenkasi, 1995).

Types of Exchanges in Organizational Processes

Given the relatively high significance placed on interfunctional knowledge communication as a component of organizational learning, the search for ways to improve this communication is strongly warranted. In order to do so, it is important to understand knowledge communication from a product exchange perspective. Such perspective takes into consideration the exchanges of tangible (e.g., parts, materials) and intangible (e.g., information, knowledge) elements between organizational functions or roles.

Interfunctional exchanges in business processes can be seen as being of two main types—material and data exchanges. As mentioned before, there have been repeated claims, particularly since the 1970s, that we are now living in an information-driven and increasingly "symbolic" society (Toffler, 1970; 1991). To these were added claims that we are in the midst of an information explosion (Davenport & Beck, 2001) where more and more people are working in the "information sector" (Hirschheim, 1985), and that organizations have become "information organizations" (Drucker, 1989).

As a result of these repeated claims, I have been curious as to the extent to which information-bearing exchanges (i.e., exchanges of data) outweigh material exchanges in organizations. This curiosity led to several studies of business processes in a variety of organizations (Kock, 1998a, 2002; Kock & McQueen, 1996, 1998a). Those studies are generally consistent with the findings of one particular study, whose target business processes are described in Table 5.1, along with the name of the organization where each business process was located.

One of the organizations from which business process-related data were obtained was Westaflex, an international car parts manufacturer based in southern Brazil. The other two organizations were based in New Zealand. One was Waikato University, whose main campus was based in the city of Hamilton,

Table 5.1. Organizational processes studied

Business Process Description	Organization
Product design	Westaflex
Parts manufacturing	Westaflex
Order delivery	Westaflex
Raw material purchase	Westaflex
University course preparation	Waikato University
University course teaching	Waikato University
Communication of a pest/disease outbreak	MAF Quality Management
Quality management consulting	MAF Quality Management
Quality inspection of parts/materials	Westaflex
Plant machinery maintenance	Westaflex
Equipment adaptation for new product	Westaflex
Software support for users	MAF Quality Management
Internal newspaper editing	MAF Quality Management
IT users support	MAF Quality Management
Staff training and development	MAF Quality Management

Figure 5.1. Distribution of exchanges according to their type

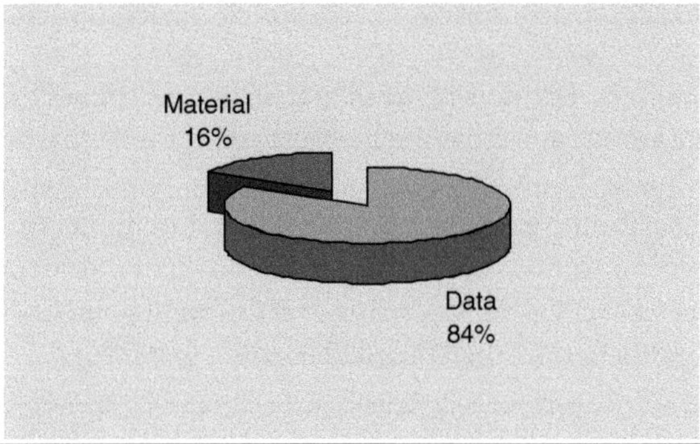

and the other was MAF Quality Management, a semi-autonomous branch of the New Zealand Ministry of Agriculture and Fisheries with offices spread throughout the country.

The study of the business processes involved the identification of data and material exchanges between the organizational functions that performed activities in each business process. Overall, 123 exchanges were identified. One hundred and three of those exchanges were found to be data exchanges, which amounts to approximately 84%. Only 20 of those exchanges, or approximately 16%, were found to be material exchanges. The distribution of those exchanges according to type (i.e., data or material) is illustrated in Figure 5.1.

The number of data exchanges was over five times that of material exchanges. Although the sample of business processes analyzed was small (i.e., 15), the large contrast between the quantity of data and material exchanges and the fact that the sample was obtained from three different organizations suggests that such contrast may be found in other organizations; an assumption that was strongly supported in follow-up studies conducted in the US (Kock, 2003; Kock & Murphy, 2001). Moreover, nearly half of the business processes studied came from Westaflex, a manufacturing organization. Thus, one can reasonably expect the proportion of data exchanges to be even higher in organizations outside the manufacturing sector. This expectation also was strongly supported by follow-up studies in the US (Kock, 2003; Kock & Murphy, 2001).

Business Process Improvement and Knowledge Communication

In previous chapters, I proposed the idea that data can be seen as carriers of information and knowledge. I also pointed out that information is predominantly descriptive, whereas knowledge is predominantly associative. Information allows us to describe the world through assertive statements about it, such as "Today it is going to rain." Knowledge allows us to associate different assertions that can occur at the same time or at different points in time, such as, "If today it is going to rain, then the road is going to be wet."

I also observed earlier in this book that business process improvement has been at the heart of what is often referred as organizational learning, particularly regarding the concept of double-loop learning proposed by Argyris (1977, 1992). However, hard evidence that empirically suggests that business process improvement is in some way causally linked with organizational learning has been very scarce, and practically nonexistent.

If we carefully look into the set of business processes described in the previous section, evidence emerges that can provide the basis for showing that business process improvement can lead to knowledge sharing. For this to happen, however, a comparison group (also known as a "control group" in experimental research methods jargon) of improvement business processes is needed. The reason is simple: the 15 business processes studied are all routine core and support business processes in the organizations from which they were taken. None of them is an instance of an improvement business process (i.e., a business process whereby business process improvement occurs).

Improvement processes are usually group-based business processes whereby business processes (although this sounds a bit confusing) are analyzed and redesigned so that some form of organizational improvement is accomplished. An example of an improvement process is MetaProi, which stands for Meta-Process for Process Improvement. MetaProi is called a *meta*-process because it is itself a business process, and yet it is used for improving other business processes[2]. MetaProi is described in detail later in this book.

Since none of the business processes described in the previous section is an improvement process, a comparison between these business processes and a new set of improvement processes could shed some light on the different nature of improvement and routine business processes regarding knowledge and information communication. It would have been even better if the new set of

Table 5.2. Improvement processes studied

Business Process Description	Organization
University course improvement	Waikato University
Undergraduate academic support improvement	Waikato University
Student computer support improvement	Waikato University
Student assignment handling improvement	Waikato University
International graduate student support improvement	Waikato University
International student adaptation support improvement	Waikato University
Software support improvement	MAF Quality Management
Newsletter editing improvement	MAF Quality Management
Pest/disease outbreak communication improvement	MAF Quality Management
Quality management consulting improvement	MAF Quality Management
IT users support improvement	MAF Quality Management
Staff training and development improvement	MAF Quality Management

improvement processes had been obtained from the same three organizations mentioned in the previous section, so organizational culture and other organization-specific factors could be eliminated as possible sources of bias in the analysis.

I was unable to obtain data from improvement processes at Westaflex, but I was able to collect data from the other two organizations. I analyzed 12 improvement processes (or meta-processes) conducted at MAF Quality Management and Waikato University by business process improvement groups, and compared these with the 15 routine core and support business processes described in the previous section. A description of these business processes is provided in Table 5.2. Groups conducted business process improvement activities according to MetaProi's guidelines.

Different communication channels have been used during the discussions. Group communication took place predominantly through an e-mail conferencing system (developed by me using Novell Groupwise[3] macros), face-to-face meetings, and phone conversations. The e-mail conferencing system allowed members to send e-mail messages to a central mailbox, which then automatically distributed the messages to all the other members of the group (e.g., as in Internet e-mail lists, also known as Listservs). Typically, group members were physically dispersed, either in different offices of the same building or campus, or across different cities or campuses. Most group members were drawn from the same business process team; that is, they were involved in the execution of the same (typically cross-departmental) business process.

I analyzed discrete and written exchanges of data (e.g., forms, reports, e-mail messages, memos, faxes, etc.) in both groups of business processes (i.e., routine and improvement processes). In doing so, I was particularly concerned with identifying information and knowledge occurrences in each of these exchanges. I did so consistently with the operational definitions of knowledge and information discussed previously in this book. A given data exchange was seen as carrying information if it had at least one purely descriptive statement, or if at least one purely descriptive statement could be extracted from it. Such statements were identified by the presence of isolated object-verb-attribute sequences, without any associative (either causal or correlational) reference. Examples of these include: "John is in Singapore this week," "Our cycle time has increased 20% in comparison with the same quarter last year," and "Our sales figures always go down at this time of the year"[4].

A given data exchange was seen as carrying knowledge if it had at least one associative statement that could be expressed as an if-then statement. Associative statements are those that associate different pieces of information in a causal or correlational way. These associations can be more or less general (or specific). General associations express knowledge in a relatively generic way. For example, consider the following statement: "I think that increased instances of litigation have been caused by our lack of understanding of our customers' needs." This statement carries knowledge that is relatively generic, because it associates two classes of phenomena—"increased instances of litigation" with a "lack of understanding of customer needs." Specific associations express knowledge in a relatively specific context. Consider this statement: "The reluctance of our chief operations officer was the main reason why our reengineering project failed." This is a much more specific statement insofar as it associates two specific instances of phenomena, namely "the reluctance of our chief operations officer" (a specific organization function) and "the failure of our reengineering project" (a specific instance of reengineering project).

The absolute counts of knowledge and information exchanges across different business process types (i.e., routine and improvement processes) are shown in Figure 5.2. There are a few important points to be made based on these aggregate numbers.

- There were considerably fewer data exchanges that carried knowledge in comparison with those that carried only information.
- Almost all data exchanges carried information. Only two data exchanges in improvement processes did not to carry any information (these ex-

Figure 5.2. Knowledge and information flowing across different types of business processes

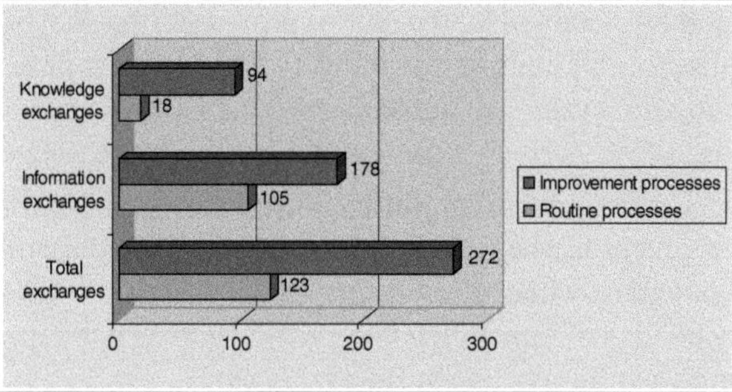

changes did not carry any knowledge, either). That is, only two data exchanges were "empty" exchanges in that they carried nothing that could be interpreted by their recipients (e.g., nonsensical exchanges such as "you may be right or wrong," or "today we tomorrow not"). This is an interesting phenomenon that also makes sense, since at least "something" (i.e., some knowledge of information) is meant to be transferred in data exchanges; otherwise, one would expect that they would not have happened in the first place.

- The number of knowledge-bearing exchanges in improvement processes (94) was much higher than in routine processes (18), in spite of the fact that more routine than improvement process were analyzed. Remember that 15 routine processes were compared with only 12 improvement processes.

Even though the number of knowledge-bearing exchanges in improvement processes was higher than that found in routine processes, a careful analysis has to account (or control) for the fact that there were more data exchanges in improvement processes than in routine processes. That is, business process improvement group members exchanged more data than members of teams performing routine business processes.

Since there were more data exchanges in improvement processes, it is reasonable to expect that there would be more knowledge exchanges, even if the ratio between knowledge-bearing exchanges and total exchanges had been the same for improvement and routine processes.

In this sense, the comparison of the figures 94 (obtained for the improvement processes) and 18 (obtained for the routine processes) may not be a totally fair one. Nevertheless, this comparison indicates that business process improvement leads to an increase, in absolute terms, in knowledge exchanges in organizations.

What is needed to complete a careful analysis of the evidence from the routine and improvement processes is to find out what the percentage of knowledge-bearing exchanges is when all exchanges are considered. This standardized figure gives an idea of how much knowledge there is per data exchange in improvement as well as routine business processes.

Figure 5.3 shows the standardized knowledge content per data exchange for improvement and routine business processes, which was obtained by dividing the number of knowledge-bearing exchanges by the total number of data exchanges for each business process type. The percentages—34% and 14%—suggest a much higher knowledge content in data exchanges taking place in improvement processes than in those observed in routine processes.

In the face of the analysis of knowledge and information exchanges, one may argue that no conclusions can be made about knowledge and information communication patterns. The reason is because I cannot be sure that the knowledge and information extracted from the data exchanges was actually

Figure 5.3. Standardized knowledge content per exchange across business process types

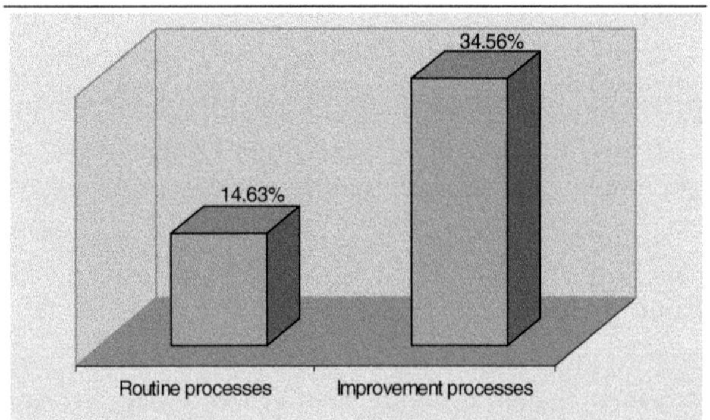

transferred between the originator of each data exchange and its recipient. In other words, the fact that knowledge and information are apparently being exchanged (according to my analysis) does not mean that they are being actually communicated.

My answer to this issue is based on one important and, in my view, sound assumption, which is that decreases in what I call "communication fidelity" should occur uniformly across the sample of exchanges analyzed.

What I mean by communication fidelity is the ratio between what is exchanged and what is really communicated. Communication fidelity can be reduced by many factors, such as the following:

- Different types of "noise"; that is, extraneous elements that distort the meaning of what is being communicated
- Lack of a shared understanding of the language used for communication
- Lack of interest in the topic about which information or knowledge is being communicated

My point is that if communication fidelity is reduced, it is reasonable to expect that such reduction would take place in more or less the same way throughout the routine and improvement processes studied. Since my conclusions are based on the comparison of figures obtained for each of these types of business processes, they should be uniformly affected by a possible reduction in communication fidelity. Such reduction could be seen as a multiplicative factor—*f*—that would be equally applied to improvement and routine process variables. Whenever relative considerations are made, *f* would automatically be cancelled out.

One of the main conclusions that can be inferred from the results presented here is that business process improvement fosters knowledge communication. Since knowledge communication is an important component of organizational learning, then it follows that business process improvement fosters organization learning.

Even though these findings are based on a small set of business processes that have been obtained from organizations in Brazil and New Zealand, they are, as previously mentioned, consistent with several other studies of business processes in a variety of organizations, including organizations in the US and other countries (Kock, 1998a, 2002, 2003; Kock & McQueen, 1996, 1998a; Kock & Tomelin, 1996).

Summary and Concluding Remarks

Knowledge is predominantly stored in organizations by means of individuals working in business process teams; that is, the organizational knowledge construct is primarily an aggregate of the knowledge held by the individuals who work in the organization (even though knowledge can be stored in documents, databases, and software). Given the increasing volume of existing knowledge in all areas, knowledge holders are pushed into specialization; that is, they are pushed into focusing on a specific body of knowledge. This leads to knowledge fragmentation in organizations, which, in turn, has been attacked by organizational learning advocates as a key reason for low business process productivity and quality. Many organizational learning proponents thus have focused their efforts on finding ways to stimulate interfunctional knowledge sharing.

A relatively new and unorthodox approach to promote interfunctional knowledge sharing is to have workers participate in business process improvement groups. While early contributors of the organizational learning and total quality management movements have hinted at the idea that business process improvement may directly or indirectly lead to knowledge sharing, there was virtually no empirical evidence pointing to business process improvement as a knowledge sharing catalyst.

This chapter discusses evidence that suggests that the number of knowledge-bearing communication exchanges in improvement processes is much higher than that observed in routine processes. It also shows that the proportion of knowledge content in communication exchanges in improvement processes is approximately 35%, compared to approximately 15% for routine business processes. These findings are particularly significant because business process improvement, unlike traditional knowledge transfer activities like training sessions and committee meetings, has other numerous side effects that are obviously beneficial. Perhaps the most important among them is business process improvement itself.

One important conclusion based on the discussion presented in this chapter is that business process improvement efforts may be employed as catalysts for organizational knowledge dissemination. This conclusion may be put into practice to support key business efforts that are becoming increasingly common, such as knowledge transfer between subsidiaries of the same company whenever new technologies and/or methods are developed; parent and acquired businesses in post-merger situations; main company and its contractors in strategic outsourcing partnerships; and main franchiser and its franchisees in large and geographically-distributed franchise chains (e.g., McDonald's and Krispy Kreme).

Even outside of the situations above, where knowledge dissemination is an important goal, business process improvement may be employed to indirectly improve business process efficiency and effectiveness by creating the opportunity for business process team members to exchange process-relevant knowledge. This assumption underlies one important organizational benefit in connection with promoting business process improvement efforts: Even if the targeted business processes are not actually changed, it is likely that process improvements will happen anyway. Those improvements will occur because the extra knowledge shared by business process team members will allow them to conduct their activities in a more integrated, better coordinated, and more cost-effective way.

Endnotes

1 This statement is often attributed to John Kenneth Galbraith. Born in 1908 in Iona State, Ontario, Galbraith is an internationally acclaimed (and sometimes criticized) economist and scholar whose seminal theoretical work has been influential in several areas of organizational research, including the incipient field of information systems.

2 In the same way as the terms meta-language and metadata are used, respectively, to refer to a higher-level language used to talk about other languages, and to higher-level data about other data sets.

3 Novell Groupwise was a leading commercial groupware product distributed by Novell Corporation at the time the study was conducted.

4 Note that the last statement is a generic statement that expresses a generalization—"something will always happen at a given time of the year." Although this type of statement may be seen as knowledge, because it allows for prediction of the future (even though this is done independently of information about the present), it is inconsistent with my

operational definition of knowledge discussed earlier in this book. Since in research it is a good practice to stick with operational definitions, statements such as this are not counted as knowledge in my analysis. Nevertheless, as I will show later in this chapter, this consideration had no effect on the main findings of the analysis.

Chapter VI

The Effects of E-Collaboration Technologies on Groups

Why Distributed Improvement and Learning?

The business environment since the 1990s is characterized by a tremendous explosion in the amount of information flowing within and outside organizations. Information flows internally between organizational functions (or organizational roles, usually distinguished by different job titles). Information flows outside organizations when communication takes place between the organization and one of its suppliers or customers.

As discussed in previous chapters, one of the main reasons why such an information flow explosion is taking place is the specialization of knowledge. As more and more knowledge is produced on a global scale, the scope of knowledge that is possessed by individuals becomes increasingly more specialized. Individuals strive to hold in-depth knowledge in a very limited number of fields and subjects; in other words, they specialize. Specialization seems to be, to a large extent, an involuntary phenomenon, which follows from human cognitive limitations and, perhaps most importantly, time constraints.

In the current business environment, those who do not specialize in one or a small number of fields tend to become less competitive in professional terms, because they do not have the time to acquire the knowledge and skills needed to compete with others who specialize in specific fields of knowledge.

If this conclusion sounds a bit too immoderate to you, try to think of anyone who could be a top criminal lawyer and, at the same time, an internationally renowned brain surgeon. Even if we are talking about a super-genius here, time constraints will prevent this from happening, as both specialties require years and years of study and focused practice to be properly performed.

However, as the number of different knowledge specialties increases, so does the need for organizations to hire and manage groups of experts who specialize in different subject areas. A large and diversified financial services firm, for example, may have to maintain hundreds of experts who specialize in different financial services. Each of these services (e.g., mutual fund management and securities analysis) are carried out by divisions that are made up of dozens of experts who specialize in different economic sectors and industries (e.g., Asian government bonds and domestic high-tech stocks). The existence of such knowledge variety leads organizations into a high degree of departmentalization (Hunt, 1996; Kock & McQueen, 1996, 1998a) or the organization around a heterogeneous structure of work teams (Eason, 1996) to cope with the management complexity that it generates.

Previous studies show that a high degree of knowledge specialization and the resulting high degree of departmentalization correlate with an intense flow of information. My own research on this topic suggests the existence of a very strong correlation between the number of organizational functions (e.g., mutual fund manager, market analyst, computer programmer, securities trader) in a business process and the number of information exchanges in it[1]. That is, the trend towards knowledge specialization seen today is also leading to a severe increase in the amount of information that has to be transferred in organizations.

To complicate this picture a bit, previous research has also pointed to a high correlation between knowledge and information flow[2]. That is, as the flow of information increases, so does the flow of knowledge. In fact, this seems to be caused by another interesting cognitive phenomenon. There appears to be an information exchange threshold above which knowledge needs to be exchanged, as well. The existence of such a threshold can be intuitively understood through the observation of the communication that takes place between pairs of workers engaged in a common business process. For example, let us consider two people engaged in the process of developing a new toothbrush, each of them being an expert in their own field (e.g., plastic materials resistance and oral preventive medicine). At the beginning of their interaction, these two people exchange descriptive information so that each of them can reach their own conclusions about their plans for new toothbrush features. However, at a

Figure 6.1. Information exchanges often lead to knowledge exchanges

John and Mark are working on the development of a new toothbrush. In order to do so, they have to exchange information and knowledge. The dialogue below illustrates an initial exchange of information (two first paragraphs) that leads to the need for an ensuing exchange of knowledge (last paragraph).

"John, you told me that the elasticity of the middle section of our toothbrushes will decrease next year. Why is that?"

"Mark, you're always the last to know things around here, aren't you? It is because we will be using high density polyethylene to manufacture them, instead of the softer low density polyethylene that we use today?"

"What? John, can you explain this to me, please?"

"Well, high density polyethylene is a very strong and hard type of plastic. If we use this type of plastic in our toothbrushes, their middle sections will be much less elastic than they are now."

certain point, they will start transferring information that does not make absolute sense to each other (i.e., information that cannot be processed based on the existing body of knowledge that each of these experts possesses). At this point, they will have to start exchanging knowledge (see Figure 6.1).

John and Mark are working on the development of a new toothbrush. In order to do so, they have to exchange information and knowledge. The dialogue below illustrates an initial exchange of information (two first paragraphs) that leads to the need for an ensuing exchange of knowledge (last paragraph).

The need for transferring increasing amounts of information and knowledge in organizations has been compounded by (or combined with) another set of trends. These trends refer to the increasing geographic independence of organizations, the increasing distribution of specialized knowledge around the world, and the spread of capitalism throughout the world.

First, no longer do organizations need to rely on local endowment factors (e.g., natural resources, cheap labor) to compete globally. Competitive advantage is now defined by the ability of organizations to acquire and deploy business process-related knowledge (Porter, 1980, 1985). A good example of this is provided by Japanese automakers like Toyota and Mazda, which managed to successfully compete in the global market in spite of Japan's lack of natural resources and a relatively expensive labor force. Compare the success of Japan with that of other countries like Brazil, for example, whose natural resources relevant to the automobile manufacturing industry have always been much more abundant and whose labor force has almost always been cheaper. It can clearly be seen that global industrial success has very little to do with country endowment factors.

Second, business process-related knowledge is increasingly found in a geographically dispersed way. People with expertise in business processes like pharmaceutical research, for example, can be found in places as far as California, Oregon, Pennsylvania, New Zealand, and Uruguay.

Third, the capitalist principles of free market and competition increasingly have been finding widespread adoption around the world, particularly since the early and mid-1990s when major free trade agreements such as the NAFTA and Mercosur agreements were signed. This has intensified competition among organizations in the same industry at a global level.

A subproduct of the above trends has been that organizations are increasingly moving towards what some refer to as the virtual organization paradigm (Davidow & Malone 1992; Mowshowitz, 1997). Virtual organizations produce and deliver their products independently of their physical location and structure. Their most important assets are not material, tangible assets, but knowledge assets. Many knowledge-intensive business processes rely largely on the transfer of information and knowledge, which can be done through the transfer of data. More and more, data transfer relies heavily on computer networks.

It should come as no surprise, then, that the higher the degree of virtuality (or virtualness) of an organization, the more likely it is that it will rely on computer networks to support communication among its members. Local and wide-area computer networks have the potential to support the acquisition, transfer, storage, and use of geographically dispersed business process-related knowledge and information.

However, as with more traditional organizations, business process teams in virtual organizations also have to cope with process inefficiencies and the need to share business process-related knowledge. Given the distributed nature of such business process teams, it becomes increasingly important that business process improvement and knowledge sharing in virtual organizations be conducted in a distributed and asynchronous (i.e., time-disconnected) manner. Hence, it is important to understanding the effects of e-collaboration technologies, particularly those that provide asynchronous group collaboration support, on business process improvement groups. After all, business process improvement groups, as I already have shown, can be a powerful tool to achieve both business process improvement and knowledge sharing.

Earlier in this book, I discussed the link between business process improvement and knowledge sharing based on the analysis of a number of business process improvement groups conducted at two New Zealand organizations, namely Waikato University and MAF Quality Management. In this chapter, I will discuss the effects of a type of asynchronous group support technology on those business process improvement groups, as well as on similar groups conducted in the US—the latter are discussed at the end of this chapter.

As it was previously mentioned in this book, the technology used to support group communication was e-mail conferencing. The e-mail conferencing system allowed members to send e-mail messages to a central mailbox, which then automatically distributed the messages to all the other members of the group (e.g., Internet e-mail lists, Listservs). Most groups had members who were physically dispersed, either in different offices of the same building or campus, or across different cities or campuses. Most group members belonged to the same business process team; that is, they were involved in the execution of the same (typically interdepartmental) business process, even though they were usually not co-located.

To facilitate understanding, I initially split technology effects into a few main categories in this chapter. The first category of effects are those on the efficiency of business process improvement groups; that is, on the organizational cost of business process improvement groups and on the total number of simultaneous business process improvement groups that an organization can possibly have at any given time. The second category of effects refers to the impact on the quality of the outcomes generated by the business process improvement groups (i.e., the quality of business process improvement proposals. The third category relates to the effects on learning or knowledge sharing effectiveness, as perceived by business process improvement group members. Each of these categories of effects is individually discussed. Following those discussions are the results of follow-up studies, partly conducted in the US.

Efficiency Effects from a Group Perspective

John Grinder[3] was the national manager of the training and certification arm of MAF Quality Management, an organization whose main goal was to ensure that certain New Zealand products, particularly food-related products, complied with high quality standards.

John was responsible for ensuring that the field inspectors had the training and the certification credentials to do their job, as required by government regulations. Since MAF Quality Management had many offices spread throughout New Zealand (made up of two islands, the North Island and the South Island), John had to rely on local centers to handle training and certification sessions. These sessions involved government-accredited consulting and training firms, as well as international certification bodies and MAF Quality Management's field inspectors. Many of those inspectors routinely audited the operations of meat and milk farms, packaged food manufacturers, and livestock enhancement companies, among other food-related organizations.

Each local training and certification center independently developed spreadsheet-based applications to keep track of information about field inspectors and the suppliers of training and certification services. However, due to recent changes in government regulations, John needed to generate periodic reports of the status of training and certification of inspectors nationwide. Moreover, John believed that those training and certification services could be partially delivered online to MAF Quality Management through the use of computer networks. John believed that this could potentially improve the quality of training, by allowing full-time access to training material by instructors, and at the same time reduce its cost to MAF Quality Management. The bottom line was that John wanted to improve the business process he managed.

John could simply hire a team of consultants to redesign the whole business process and develop a distributed computer system to integrate data from the several local centers scattered throughout the country. But John knew better than that. If he wanted the new business process to work, John knew that he would have to explain to the local centers why it was necessary to change the way training and certification was carried out. He also knew that he would have to give them the opportunity to propose business process changes themselves. It was either this, or the local centers would resist any change. "People don't dislike change; people dislike change being imposed on them," John said to himself.

The problem was that in order to get everyone's input regarding the business process change, John would have to bring together a group of at least 20 people representing each of the various local centers. And he was not sure his quarterly budget would allow him to pay for the travel and accommodation expenses of all those people, not to mention buyout fees that some offices wanted to charge him to compensate for the hours the staff would spend in the business process improvement effort. MAF Quality Management's administrative structure was highly decentralized, with many departments independently handling their own budgets.

John eventually decided to run the whole discussion as a business process improvement group whose members would interact through an e-collaboration system. He contacted each of the local centers by phone and invited 23 people to participate in the e-collaboration technology-supported business process improvement group.

The discussion lasted about three months, and the whole group met face-to-face only once at the end of the discussion. John estimates that the use of the e-collaboration system has saved him over $60,000 in just travel and accommodation expenses. Also, he did not have to pay for the staff-hours spent in the e-collaboration technology-supported discussion, since those staff did not have to physically leave their offices during the discussion.

Moreover, John was pleased with the results. Representatives of local centers proposed and, after some discussion, almost unanimously agreed on adopting a new set of procedures. They went some way towards specifying software and hardware requirements for a computer system to enable the new procedures, which was soon after implemented by a software firm.

The Organizational Cost of Improvement Groups is Reduced

Although the picture painted above may seem a bit too rosy, it is true to the general trend observed in the business process improvement groups I have facilitated and studied in the past. However, a critical analysis of those business process improvement groups suggests that not all groups were as successful as the group led by John. Some of the groups studied, for example, failed to reach a consensus about business process changes, and, as a result, no changes were proposed or implemented by these groups. A nutshell discussion of business process improvement group success factors is provided later in this book. Some considerations regarding the general quality of e-collaboration technology-supported business process improvement group outcomes are made later in this chapter.

Nevertheless, one important trend that was observed in the 12 e-collaboration technology-supported business process improvement groups conducted at

Waikato University and MAF Quality Management is their reduced cost compared with similar face-to-face business process improvement groups. Group members indicated in interviews that such cost reduction took place. They were asked to compare the e-collaboration technology-supported business process improvement groups with previous face-to-face business process improvement groups in which they had participated.

Twenty-nine out of 46 (or 63%) of the business process improvement group members at Waikato University spontaneously remarked that e-collaboration technology support had reduced the costs of member participation in business process improvement groups. Two main reasons for this cost reduction were mentioned. A reduction in disruption costs (i.e., costs related to employees having to interrupt their routine activities in order to participate in the business process improvement groups) was one reason given by most of the interviewees. The other reason was a reduction in the time each member spent participating, either actively or passively, in the group discussion.

When asked about a possible reduction in travel and accommodation costs, all business process improvement group members stated that those cost reductions were very significant in most groups, but too obvious to be mentioned in the interviews. The following quote is from a faculty member who had been involved in a business process improvement group dealing with legal issues related to the provision of academic advice to students.

It is very hard [...] to organize meetings around people's schedules. It was probably a lot quicker to respond to [electronic contributions] than to get together and sit in a face-to-face meeting and talk about other things for a while until you get on with the subject at hand. [...] It probably increased the input that you got from other departments.

At MAF Quality Management, structured interviews indicated that approximately 78% of the respondents viewed e-collaboration technology-supported business process improvement groups as having cost the organization much less than face-to-face business process improvement groups. This is illustrated in Figure 6.2, where the distribution of response frequencies suggests a statistically strong trend[4] in the direction of a collective perception that e-collaboration technology support leads to a reduction in the overall cost of running a business process improvement group.

Figure 6.2. Perceptions about e-collaboration technology support impact on group cost

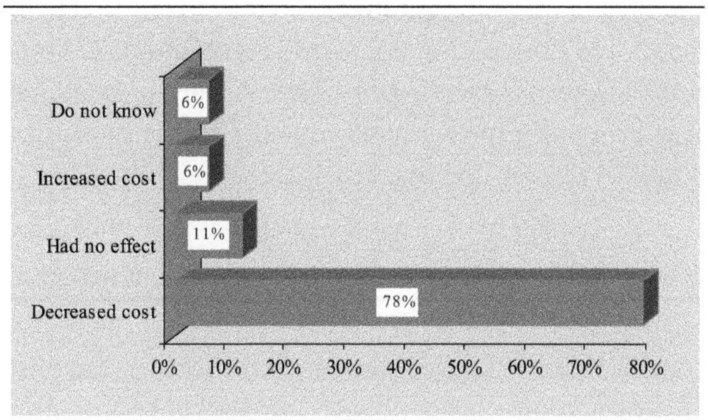

The average time spent by a member of a business process improvement group discussion at MAF Quality Management was estimated to be slightly over 20 hours, if the discussions was carried out exclusively through face-to-face meetings. This time appears to have been reduced by e-collaboration technology support to about one hour and 30 minutes, which was the average amount of time spent by ordinary business process improvement group members (i.e., all group members, with the exception of the group leader). This amounts to an average reduction of approximately 93%.

For group leaders, the average time spent in the group discussion went down from the estimated 20 hours to approximately seven hours, according to direct time measurements. Although less than the average time for ordinary members, this amounts to a 65% reduction in the group leader participation time.

While the above time reductions may look a bit suspicious at first glance, or too good to be true, they make more sense when one considers all of the time-wasting activities normally involved in face-to-face meetings (e.g., fitting the meeting in one's schedule, transporting oneself to the venue of the meeting, socializing, discussing tangential issues, etc.).

The bottom-line money saving per group was gauged through an absolute dollar amount estimate, which proved to be very attractive. To obtain such an estimate, I assumed very conservatively that a group member costs the organization on average $15.00 per hour. Since the average number of members in a business process improvement group is nine (averaged based on the business process improvement groups at MAF Quality Management), then e-collaboration technology support reduced the cost per group to the organization at least $2,415.00. This is the price of a good laptop computer with basic office automation software installed on it. Note that group facilitation (provided by a business process improvement consultant, for example) and travel expense savings have been disregarded in the calculation of this figure.

The Lifetime of Improvement Groups is Reduced

Another trend observed in the e-collaboration technology-supported business process improvement groups conducted at Waikato University and MAF Quality Management was an apparently shortened lifetime, compared to face-to-face business process improvement groups. And yet, as it will be seen later in this chapter, this reduced lifetime has not led to a decrease in the quality of group outcomes (i.e., the business process redesign proposals generated by the groups). Business process improvement group lifetime was measured in days, from the inception of the group to its formal cessation.

The lifetime of the business process improvement groups conducted at Waikato University varied from 32 to 54 days, with an average of about 40 days, and a standard deviation of approximately nine days. At MAF Quality Management, business process improvement group lifetime was slightly shorter. It ranged from 10 to 29 days, averaging 22 days, with a standard deviation of approximately eight days.

At Waikato University, many business process improvement group members voluntarily and consistently noted a perceived decrease of group lifetime as a consequence of e-collaboration technology support, though no frequency distribution analysis was performed. At MAF Quality Management, interviewees' responses displayed a strong trend towards the perception that e-collaboration technology-supported business process improvement groups had been completed in less time, measured in number of days, than face-to-face business process improvement groups. Approximately 78% of the interviewees were of this opinion, which indicates a statistically strong trend[5]. This is illustrated in Figure 6.3, which, coincidentally, is very similar (but not identical) to Figure 6.2.

Figure 6.3. Perceptions about e-collaboration technology support impact on group lifetime

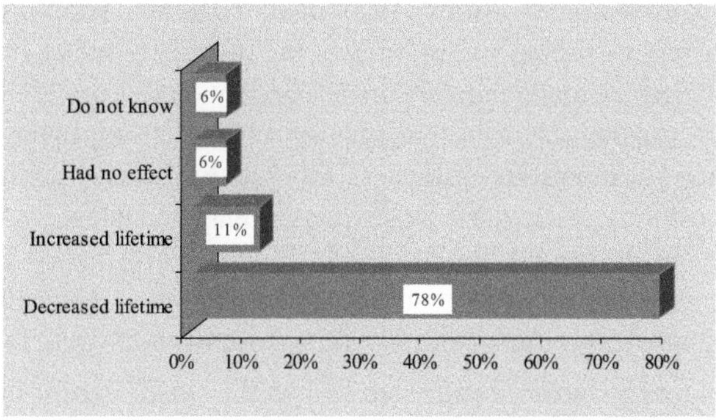

The main cause for the reduction of group lifetime, according to group members, was a reduction in what many referred to as the group setup time. This was generally described as the time needed to accomplish group setup activities, such as defining a list of problems (or improvement opportunities) to be discussed by the group, selecting group members, and inviting those members to take part in the group. Many interviewees noted that while the time needed to carry out some of these activities had been curtailed by e-collaboration technology support, the need for other activities more typical of face-to-face meetings had been completely eliminated. Examples of such activities are choosing and preparing a venue for the group meetings, and coordinating member attendance.

Improvement Initiatives are Decentralized

Managers have long dominated the scene when it comes to changing their organizations. The total quality management tried to turn this picture around a little by showing that the best prepared to improve business processes were those who carried them out, not those who managed them. Yet, in spite of this, business process improvement has long suffered from an over-reliance on managers as far as business process change decisions are concerned. And the business process reengineering movement, with its top-down change philosophy, contributed towards legitimizing this situation.

However, managers are just a few when compared to the total number of employees in most organizations. And with the current corporate downsizing trend, the manager-to-employee ratio is being steadily reduced, day after day. One self-proclaimed management guru has been quoted as proudly saying that, "where years ago there was one manager for each seven employees, now there is one for each one hundred employees." Obviously, this is an achievement for many organizations, particularly those where management levels are more of an information buffer than anything else (i.e., they pass information up and down the organizational hierarchy, without adding anything to it). After all, managers, like other support and control entities, do not usually add value to customers.

But as managers grow more and more scarce, they also become busier. Their endless stream of business meetings and improvised interactions (Kurke & Aldrich, 1983; Mintzberg, 1975) becomes more fragmented, and the likelihood that they will want to or have the time to participate in more meetings grows increasingly slim. Such an environment, though purportedly more cost-effective in terms of management expenses and utilization, is not a fertile ground for business process improvement ventures. Managers, whose leadership and seniority are needed to legitimize and give weight to business process improvement efforts, are hardly available.

Apparently, e-collaboration technology support reverses this picture by reducing the demand for senior leadership in business process improvement groups. At Waikato University, some business process improvement group leaders spontaneously remarked that e-collaboration technology support had made it considerably easier for them to lead their business process improvement groups. None of those leaders were the most senior members of their business process improvement groups. One of those group leaders, the most junior person in his group, stated:

> *... leading a face-to-face meeting [with the same group members] would be considerably more demanding and stressful for me.*

According to the perceptions of the business process improvement group leader that provided this statement, his junior status in the organization likely would have hindered him considerably from leading the business process improvement group, had the group been conducted only through face-to-face discussions.

At both Waikato University and MAF Quality Management, similar reasons were put forth by business process improvement group members for the reduction of in-group demand for leadership seniority apparently associated with e-collaboration technology support. Following are the most common reasons:

- A suppression of social cues by the electronic medium, which could differentiate junior from senior members.
- A reduction in the influence that individual members have on the group, which was seen as likely to increase with seniority in face-to-face groups.
- A suppression of hierarchy barriers to an open discussion.

Note that these reasons almost imply a sense of partial anonymity in discussions conducted through an electronic communication medium. Yet, none of the business process improvement groups involved anonymous electronic contributions at all. That is, the electronic postings that were exchanged by business process improvement group members were not anonymous; the names of the senders were shown at the top of their postings, as usually is the case with e-mail. Yet, the perception of partial anonymity was very much a factor in the communicative behavior displayed by business process improvement group members. The following quote from a group leader, who refers to the e-collaboration system as simply "e-mail," illustrates the member perceptions underlying their explanations already listed.

Normally if I am in a [face-to-face] situation and with [another member's name - removed], who is my boss, his opinion counts over mine, when I'm sitting in the same room ... on e-mail I feel just as equal—I don't feel that he will influence me or that his opinion will be more important than mine. Because I feel like I can just freely put my ideas on an e-mail and I don't feel threatened by him being above me. You [referring to business process improvement group members in general] are all equals on e-mail ... I definitely don't think about the hierarchy structure when I'm on e-mail, but I do think about it when I'm sitting in a room and I see [names of two other group members - removed] sitting there. And they get a lot more influential because of that, because everybody is a bit more wary of what they say, whereas on e-mail people are more likely to say what they've got to say.

Figure 6.4 shows statistically significant (according to Chi-squared tests) perception trends for members polled across different business process improvement groups that provide majority support for all of the above-mentioned explanations. The figure shows the distribution of perceptions of business process improvement group members in connection with the following:

- Social cues: Refer to perceived cues (e.g., body language, dress code) that indicate differences in social and/or organizational status.
- Member influence: Refers to the influence that each member is likely to exert on other members of the group, particularly in connection with the business process redesign decisions made by those other group members.
- Hierarchy barriers: Refers to perceived barriers imposed on individual group members on their free expression of ideas due to differences in the positions that different group members occupy in the organizational hierarchy.

As shown in Figure 6.4, e-collaboration technology support significantly reduced the importance for group members of perceived cues (e.g., body language, dress code) that indicate differences in social and/or organizational status, the influence that each member is likely to exert on other members of the group, and the perceived barriers imposed on individual group members on

Figure 6.4. E-collaboration technology support impact on factors influencing group demand for senior leadership

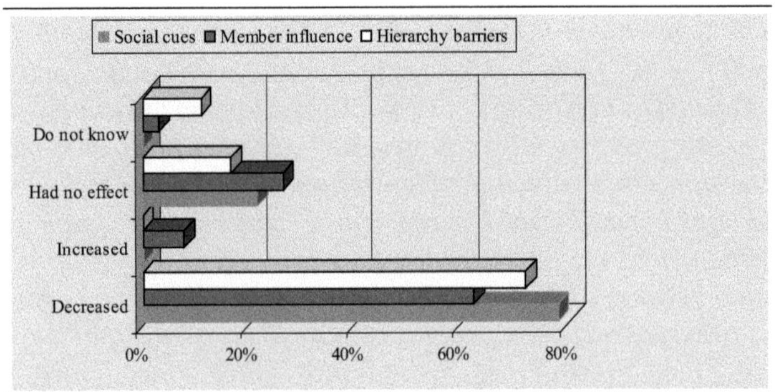

their free expression of ideas due to differences in the positions that different group members occupy in the organizational hierarchy.

The Number of Possible Improvement Groups is Increased

As demand for senior leadership is reduced by e-collaboration technology support, so is the need to rely on managers to lead business process improvement groups. Ordinary business process improvement group members, as opposed to leaders, do not need to be managers; they can be anyone within the organization or even from outside the organization. They can be external customers and/or suppliers. A direct and obvious consequence of this relaxation in leadership (and, for a variety of reasons, overall membership) requirements afforded by e-collaboration technology support is a multiplication of the number of possible business process improvement groups that can be run at a given time in the organization. To this, it can be added that the lifetime of business process improvement groups is reduced by e-collaboration technology support, making this effect even stronger. That is, the increase in organizational process improvement groups' capacities is combined with a shorter lifetime to yield a potential expansion in the number of business process improvement groups that can be conducted per unit of time (e.g., per quarter) in an organization.

These combined effects could be easily and clearly observed at Waikato University. This organization instituted official business process improvement days, in which administrative staff and faculty members were expected to engage full-time in business process improvement group discussions during an entire day. Business process improvement groups would look into current business processes and related organizational procedures and regulations, and then propose changes aimed at business process improvement. Although those group efforts were typically seen as relatively successful by most, their frequency was very low, usually twice every year. As soon as e-collaboration technology support was made available to staff and faculty, five groups were conducted within less than a quarter.

Many business process improvement group members at both Waikato University and MAF Quality Management pointed out that e-collaboration technology support had made it much easier for them to start and conduct their business process improvement discussions with a minimum of disruption for them and their fellow group members. Several of these members spontaneously mentioned a reduction of group setup time as an explanation for their perceptions.

Effects on Group Outcome Quality

Whatever efficiency gains are obtained through e-collaboration technology support, it would be difficulty to justify the use of an e-collaboration tool if it impaired the quality of group outcomes. The main outcome of any business process improvement group is the set of business process change proposals generated by the group. Such outcome is the focus of this section. Here, we are concerned with the impact that e-collaboration technology support has on the quality of business process improvement proposals in comparison with that of face-of-face groups.

Still using the business process improvement groups conducted at Waikato University and MAF Quality Management as a basis, group members were asked to compare their experience in the e-collaboration technology-supported business process improvement group discussion with that of similar face-to-face situations. An aggregate analysis of their answers indicates an interesting yet slight trend towards a perceived increase in the quality of business process redesign proposals generated by groups. This trend was

Figure 6.5. Perceptions of e-collaboration technology support impact on group outcome quality

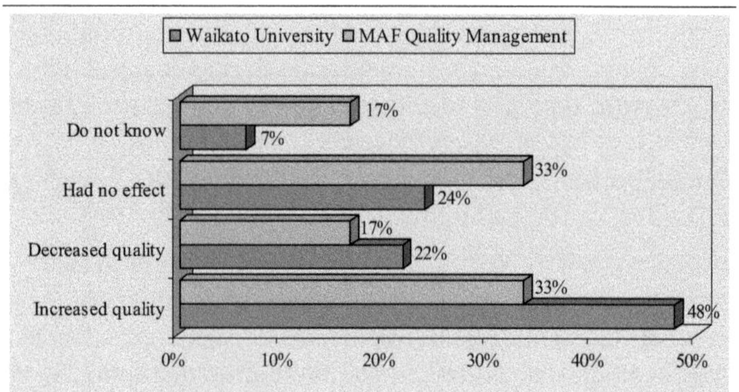

observed at both Waikato University and MAF Quality Management, and was marginally stronger at the education institution. Figure 6.5 shows the frequency distribution of answers from business process improvement group members regarding business process redesign quality in both organizations.

As in previous analyses of perception frequency distributions, I used a statistical analysis technique called Chi-squared test to establish the statistical strength of the perception trend. This test revealed a 5% probability that the trend observed at Waikato University was due to chance. For MAF Quality Management, the test yielded 63% of certainty regarding the perception trend, or a 37% probability that the trend was due to chance. In other words, if I were to generalize the findings to other areas (or departments) of either organization, I would be much more confident regarding the validity of such generalization for Waikato University than for MAF Quality Management.

A decisive trend towards an increase in connection with group outcome quality due to e-collaboration technology support cannot be inferred based on the evidence collected, as there were a considerable number of members who perceived it otherwise. Seventeen percent of the business process improvement group members at MAF Quality Management, and 22% at Waikato University, perceived a decrease in quality due to e-collaboration technology support (see Figure 6.5). Nevertheless, the perception frequency distribution in Figure 6.5 does suggest a general and statistically strong trend towards a non-negative effect on business process redesign quality. Seventy-two percent of the respondents at Waikato University perceived either an increase or a null effect in business process redesign quality; this proportion was about 66% at MAF Quality Management.

Most of the group members who perceived an increase in the quality of business process improvement group outcomes due to e-collaboration technology support explained it by an improvement in the quality of individual contributions from business process improvement group members interacting through the computer system. That is, they noticed an increase in the quality of individual contributions, which, they reasoned, had led to an increase in the quality of the business process redesign proposals generated by their groups. The remarks below, from a business process improvement group member, provide a good illustration of this general perception.

You think more when you're writing something, so you produce a better quality contribution. Take, for example, what [member's name - removed] wrote, she wrote a lot and it seemed that she thought a lot about it before she e-mailed it to the group. She wasn't just babbling off the top of her head, she tended to think out what she was writing. I know I did it a lot, specially my first message. I really thought a lot to put it together.

The majority of those who perceived a decrease in business process redesign quality believed that it had been caused by the inherent characteristics of communication conducted primarily through an electronic medium. In these members' judgments, interacting through the electronic communication medium increased the ambiguity in the business process improvement group discussion. This was particularly true, according to these business process improvement group members, in the analysis stage of a business process improvement group, where the group analyzes the target business process (or business processes) for redesign. Group members had to build a shared understanding of the business process being analyzed in this stage, so they could later effectively contribute business process change suggestions.

Most groups had a multi-departmental composition. This meant that differences in the technical language used by different members had often become obstacles that had to be removed if the discussion was to proceed successfully. The asynchronous and distributed nature of the communication supported by the e-collaboration tool in the business process improvement groups prevented immediate feedback and the use of non-verbal cues (e.g., gestures), which apparently made it more difficult, in the opinion of several group members, to remove obstacles to a shared understanding about the business process or processes being analyzed.

Still, the percentage of respondents who were of the opinion that redesign quality had been decreased by the use of group technology was comparatively low. This indicates that the higher quality observed in individual contributions might have offset the communication constraints inherent in the electronic medium. As a consequence, a neutral bending towards positive overall effect on business process redesign quality can be reasonably inferred from the evidence collected. That is, in spite of potential difficulties associated with communicating electronically, the business process improvement groups did not do much better or worse than similar face-to-face groups.

Effects on Knowledge Sharing Effectiveness

Earlier in this book, I concluded that business process improvement fosters, as a group process, knowledge communication. This conclusion was largely based on a content analysis of communication interactions between people carrying out business process improvement and routine business processes in real organizational settings. In this section, my main concern is shifted to the effect that e-collaboration technology support has on the effectiveness of knowledge sharing. In other words, do business process improvement group members interacting electronically exchange knowledge to the same extent that they do face-to-face?

As shown previously in this book, even though more knowledge communication is apparently closely linked to the task of improving business processes, it is important to establish whether e-collaboration technology support does not interfere negatively with the learning process that business process improvement group members undergo when they exchange knowledge (and information).

The perception aired by some business process improvement group members that e-collaboration technology support may increase communication ambiguity, which was discussed earlier in this chapter, is consistent with previous research, notably research associated with a very influential theory of media adoption and use known as media richness theory (Daft & Lengel, 1986; Kock, 2002a; Lee, 1994). Media richness theory argues that different communication media possess varied amounts of an abstract property called "richness," which relates to the capacity of the media to convey more or less information and knowledge.

According to media richness theory, the richest communication medium of all is that afforded by face-to-face interaction (Kock, 2002a; Lee, 1994). Media that prevent non-verbal cues from being communicated and which delay feedback, like e-mail, are seen as less rich than media which allow immediate feedback and the communication of non-verbal cues, features that are found in abundance in face-to-face meetings (Daft et al., 1987; Lengel & Daft, 1988). It follows that e-collaboration systems that enable asynchronous and distributed communication interactions provide a leaner medium for communication than do face-to-face meetings. Media richness theory argues that lean media are less adequate to the transfer of knowledge and information[6] than rich media.

Given expectations based on media richness theory, pessimistic expectations about the impact of e-collaboration technology support on knowledge sharing and learning in business process improvement groups can be reasonably assumed. Moreover, the empirical research literature on e-collaboration technologies that enable asynchronous and distributed communication has reported a number of failures of these technologies to support interdepartmental knowledge communication (Kock, 1997; Kock & Davison, 2003). Among other reasons, these failures have been explained by the following:

- The inherent ambiguity that the electronic medium adds to group communication (Rogers, 1992).
- Social norms and reward systems adopted by firms, that could themselves become obstacles to knowledge sharing (Orlikowski, 1992) through e-collaboration technologies.
- The lack of balance between the benefits to those who have to do extra work because of the introduction of an e-collaboration system and those who do not (Grudin, 1994).

The perceptions of the business process improvement group members regarding individual learning, however, contradicted the gloomy picture painted by previous theoretical ideas and empirical research. Group members were consistently positive in their views about e-collaboration technology support impact on individual learning (which is related to knowledge sharing), as shown in the distribution of perceptions depicted in Figure 6.6. As with previous analyses of frequency distribution, this trend was checked for statistical significance. This revealed that the probability that the trend observed was due to chance is about 2.5% for the frequency distribution at Waikato University and 0.5% for MAF Quality Management. That is, it can be safely assumed that the trends are not due to chance. My interpretation is that these trends are most likely a result of underlying capabilities of the technology to move group behavior towards learning-conducive situations.

A little unexpected is the explanatory direction pointed at by the evidence, particularly the evidence presented in previous sections of this chapter. There, group members perceived an increase in the quality of their group outcomes (i.e., business process redesign proposals) resulting from the use of e-collaboration technology support to their interaction as members of business process improvement groups. One of the main reasons for this, according to the

Figure 6.6. Perceptions about e-collaboration technology support impact on learning

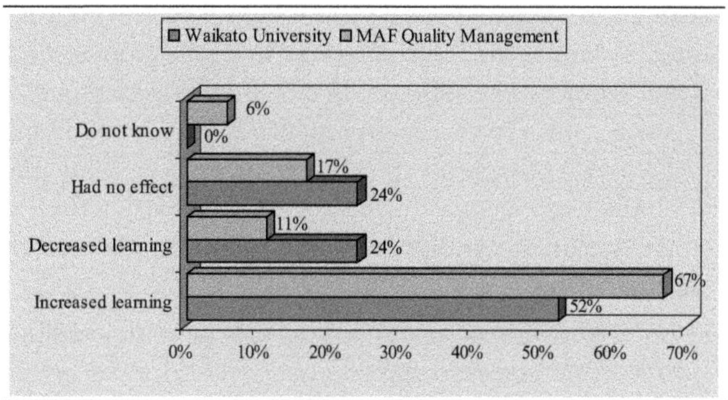

group members themselves, was the better preparation of individual contributions by group members. Yet, the same members perceived that computer mediation itself had the potential to increase communication ambiguity. A little strange, don't you think?

What is the direction at which the evidence above points in light of the perception trends shown in this section? It appears that the answer is that the leanness of the communication medium afforded by e-collaboration technology support has led to an adaptive behavior on the part of the group members that overcame, at least partially, the constraints posed by the leaner medium. More knowledge was communicated (which is consistent with the conclusions reached earlier in this book), which led to higher perceived learning.

When asked to explain their answers regarding e-collaboration technology support impact on individual learning, there was a clear convergence in the answers given by business process improvement group members. Most of the relatively small number of members who perceived a decrease in individual learning, suggested that this was due to a reduction in the degree of interaction caused by the e-collaboration technology mediation of the group communication. That is, according to that small number of members, less individual contributions are made in discussions mediated by e-collaboration tools than in face-to-face discussions, and thus less learning takes place.

The idea that e-collaboration technology mediation leads to reduced group interaction is certainly incorrect for same-time-same-place interaction (as in group interaction through traditional group decision support systems), where the number of contributions per unit of time has consistently been found to have been increased (Dennis et al., 1996; Sheffield & Gallupe, 1993). A good example of this type of same-time-same-place e-collaboration system is Meetingworks, which at the time of this writing was commercialized by a company of the same name headquartered in Seattle, Washington. Much of the pioneering research on group decision support systems was conducted at the University of Arizona by a group of researchers led by Jay Nunamaker, who has been recognized by *Fortune* magazine, the Association for Information Systems, and other media outlets and community organizations for his groundbreaking research.

Having said this, it is also important to point out that the commercial success and use of same-time-same-place e-collaboration systems, such as Meetingworks, have not even come close to the commercial success and use of e-collaboration systems enabling asynchronous (time-disconnected) and distributed communication, of which perhaps the simplest example is e-mail.

In my study of asynchronous and distributed business process improvement groups supported by e-collaboration tools, partially discussed here, group interaction, indeed, seems to have been considerably reduced by e-collaboration technology mediation. Yet, whether this led to a reduction in information and knowledge exchange is doubtful, since the length of individual contributions and their knowledge content appears to have been considerably increased. This effect may be partly due to e-collaboration technology mediation.

At MAF Quality Management, the two main reasons presented for the perceived increase in individual learning were better quality of and more sincere (or fear-free) individual contributions. Business process improvement group members linked the increase in sincerity to the sense of personal insulation fostered by the e-collaboration technology mediation. That is, often when members interact via an e-collaboration tool they feel less inhibited to freely express their feelings and ideas. This perception is partially supported by previous studies of e-mail communication in organizational contexts (Sproull & Kiesler, 1986), distributed and anonymous groups (Jessup & Tansik, 1991), and groups composed of introverts and extroverts (Yellen et al., 1995).

Business process improvement group members at Waikato University gave virtually the same explanations as those at MAF Quality Management regarding their perceptions in connection with e-collaboration technology effects on member learning. Yet, they added a new one to our repertoire, which was that member learning was improved by the higher departmental heterogeneity afforded by e-collaboration technology support. According to those members, the distributed and time-independent nature of the computer-mediated interaction allowed for a larger number of departments to be represented in each business process improvement group. This, in turn, brought into the discussion more ideas and individual perspectives that were new to many group members, thus leading to a perception of increased individual learning.

Further Evidence from Follow-Up Studies

I have conducted and coordinated many follow-up studies similar to the ones discussed in the previous sections, particularly in connection with organizations in the northeastern U.S., and more specifically in and around the Philadelphia metropolitan region. This comprises several townships in Philadelphia's surrounding region, in the states of Pennsylvania and New Jersey.

The results of those studies have generally supported the conclusions reached in the previous sections. In fact, the similarity of the findings is, in and of itself, fairly interesting. After all, most of the data discussed in the previous sections have come from the study of business process improvement groups conducted in New Zealand and Brazilian organizations.

Reaching very similar conclusions based on the study of business process improvement groups conducted in New Zealand, Brazil, and the US, which are different countries with significantly different cultures, allows us to reasonably assume that the conclusions reached so far regarding the effects of e-collaboration technology support on business process improvement groups are likely to be somewhat culturally independent. That is, those effects may be, to a certain extent, universal and related to how human beings are designed, as far as our biological communication apparatus is concerned. This is an interesting theme, which will be picked up later in this book.

Among the studies I coordinated relatively recently is the one conducted by Dorrie DeLuca (2003), which she carried out as part of her doctoral work at Temple University. DeLuca received her Ph.D. degree for that work in 2003 from Temple University's Fox School of Business and Management. Like the business process improvement groups discussed in previous sections of this chapter, the groups studied by DeLuca were typically small in size (from seven to 11 members), had a relatively short lifetime (no more than two months), during which its members defined, analyzed, and searched for alternatives to improve one or a few organizational processes. The e-collaboration technology support used by DeLuca's groups was very similar in many aspects, particularly in terms of functionality, to the e-collaboration technology support used by the groups discussed in the previous sections.

Again, similarly to the groups discussed earlier in this chapter, most of DeLuca's business process improvement groups were cross-departmental (i.e., they involved members from more than one department), and targeted cross-departmental business processes (i.e., business processes that involved more than one department in their execution). Finally, like the groups discussed in the previous sections of this chapter, most of DeLuca's business process improvement groups were "zero history" groups, in that their members in the past had not participated together in business process improvement groups; although often group members were familiar with each other.

Given the evidence from past research that communicating electronically is more difficult than communicating face-to-face, I decided to put forth a tentative way of estimating the degree of difficulty associated with communicating information and knowledge through a given medium. I argued that such estimation could be achieved through a measure that I called "fluency" (Kock, 1998). Fluency is calculated as the number of words per minute that individuals can exchange using a particular medium, and can also be seen as a measure of the amount of cognitive effort involved in the communication (Kock & D'Arcy, 2002). The more difficult it is to communicate through a particular communication medium, the lower is the fluency through that medium.

One interesting pattern in DeLuca's business process improvement groups is the range of fluencies observed. Fluencies in DeLuca's study ranged from a minimum of 1.2 to a maximum of 9.9 words per minute—these are averages per group. The average fluency for all the business process improvement groups was 4.8 words per minute, which is much lower than the approximately 116 words per minute that one would normally expect to see in face-to-face meetings discussing issues of about the same level of complexity as business process improvement groups (McQueen et al., 1999).

It is important to point out that the difference in fluencies between groups communicating electronically and face-to-face mentioned above cannot be explained based on the known fact that "typing is slower than speaking" (that is, copy-typing is slower than reading out loud a pre-prepared piece of text), as average typists have been known to be able to type 60 to 70 words per minute (McQueen, 1991; McQueen et al., 1999). Even when this typing-versus-speaking effect is controlled, we can still conclude that fluency was reduced by over 80%, apparently due to the change in communication medium from face-to-face to electronic. Such a large reduction in fluency can be explained based on the inherent cognitive difficulty associated with communicating electronically, an idea that was put forth earlier in this chapter and that is explored further later in this book.

If we use McQueen et al.'s (1999) detailed study of face-to-face groups (which performed tasks of comparable complexity to business process improvement groups) as a basis, we also can highlight key differences between those groups and DeLuca's business process improvement groups, whose members interacted primarily electronically. One of those differences is the lower number of contributions in each of the e-collaboration technology supported groups, ranging from 10 to 40 electronic postings, which contrasts with the average of 481 oral contributions in the face-to-face groups found by McQueen and colleagues. A second difference is a lower word count per group, which was about half the average word count in the face-to-face groups. A third difference is a higher number of words per individual contribution in each group, ranging from 203 to 280, which contrasts with the average of 37 words per oral contribution in the face-to-face groups. Finally, a forth difference is a much longer time to reply to electronic contributions by others, ranging from 1.3 to 3.4 days, which contrasts with the average of less than one hour in the face-to-face groups studied by McQueen and colleagues.

These differences can be seen as reinforcing the higher cognitive effort hypothesis previously discussed, which essentially states that members of business process improvement groups interacting through an electronic communication medium are generally forced to think more about their contributions before they make them than if they were in face-to-face situations. For example, the difference in reply time suggests that group members potentially reflect more on their contributions before they post them electronically than they do in face-to-face groups (where often members don't have the opportunity to think much about what they say before they say it). The difference in contribution length suggests that individual electronic contributions (i.e., electronic postings) are relatively more elaborate than individual face-to-face contributions (i.e., oral contributions in a face-to-face meeting).

The above-mentioned differences make it easier to understand why the number of contributions and total word count in e-collaboration technology supported groups seems to be lower than in face-to-face groups, since with better thought out and more elaborate individual contributions, the total amount of necessary communication is reduced, and so are the total number of electronic postings and words that need to be exchanged.

Interestingly, the mean individual contribution (i.e., electronic postings) sizes for the e-collaboration technology supported groups approached the maximum number of words fitting on a single screen, which was approximately 250, plus or minus 20 words, depending on font size and distribution of words on the screen.

The evidence regarding patterns of perceptions by business process improvement group members in connection with several variables in DeLuca's study is virtually identical to that obtained by me in the study of the two New Zealand organizations, discussed earlier in this chapter.

Figure 6.7 summarizes the results of an analysis of aggregate evidence collected from both the U.S. and New Zealand, where the number of respondents polled totaled 68, and whose responses were overall very similar. Those respondents were members of business process improvement groups, and in this particular sample of respondents, no one gave a "don't know" answer. Questions in connection with an expanded set of variables were asked. Figure 6.7 summarizes the distribution of answers along a three-point scale—increased, had no effect, and decreased.

Figure 6.7. Perceptions about e-collaboration technology effects

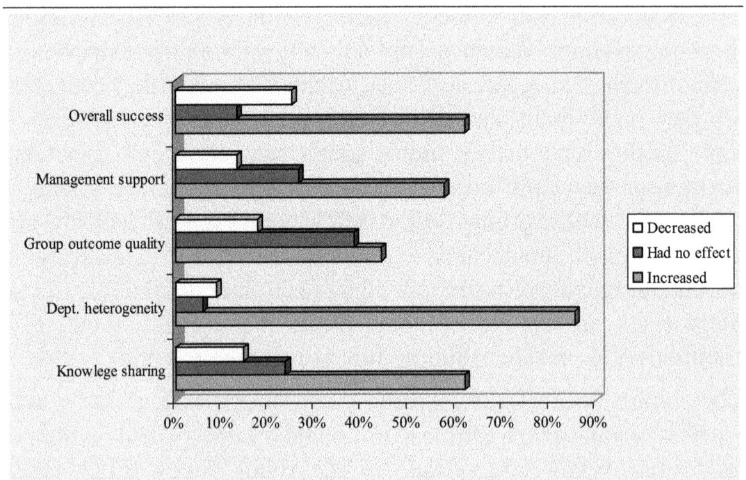

The questions were in connection with the participants' perceptions of the impact of the e-collaboration technology support (from the bottom up in Figure 6.7) on knowledge sharing among business process improvement group members, departmental heterogeneity in the business process improvement groups, group outcome quality, likely management support to business process improvement group decisions, and likely overall success of a full-blown business process improvement program using an e-collaboration technology like the one used by the groups in the respondents' organization.

The evidence in connection with the knowledge sharing, departmental heterogeneity, and group outcome quality refers to the respondents' perceptions regarding the direct impact of the e-collaboration technology support on attributes of the business process improvement groups in which they had participated. The evidence in connection with management support and overall success refers to the respondents' perceptions regarding the potential impact of the e-collaboration technology support on management support to business process improvement group decisions, and on the overall success of a full-blown business process improvement program in their organizations, respectively.

Chi-squared coefficients were calculated for each of the perception distributions shown in Figure 6.7. Those coefficients suggest that all the distribution trends toward a positive perception of e-collaboration technology support effects on the several variables are statistically significant. The likelihood that the trends were due to chance is less than 5% for group outcome quality, and less than 1% for all the other variables. In other words, there was a statistically significant trend among group members toward seeing e-collaboration technology support as having had a positive impact on knowledge sharing, departmental heterogeneity, group outcome quality, likely management support, and likely success of a full-blown project, in the context created by business process improvement groups. The reasons provided by group members to explain their answers were virtually the same as those discussed in the previous sections of this chapter, whether the answers were positive, negative, or neutral.

The patterns in this evidence provide general support for a positive outlook regarding the impact of e-collaboration technology support on business process improvement groups. Nevertheless, an assessment of actual group success would lend further evidence that would be useful in the evaluation of how aligned with reality was the perception-based evidence discussed here. Such an assessment can be conducted based on the implementation of business process redesign proposals and related business results. It is reasonable to assume that if the perceptions were poorly aligned with reality, then the level of success of the groups studied would be generally lower than that reported in the empirical literature of face-to-face business process improvement initiatives.

In our study, business process improvement group success was assessed based on in-depth unstructured interviews with group members, other employees, and managers after the business process improvement groups were completed and their process redesign recommendations were implemented. The assessment was based on the business implementation of business process redesign proposals and related business results. For the purposes of this study, business process improvement groups were seen as either successful or unsuccessful, according to criteria proposed in the process improvement literature (Burke & Peppard, 1995; Davenport, 1993; Hammer & Champy, 1993). Following those criteria, business process improvement groups were categorized as successful only if the business process changes recommended by them were implemented fully or partially and led to positive observable results; otherwise, they were categorized as unsuccessful. Also, the success rates of business process improvement groups in this study were compared with success rates obtained from a large multinational survey of traditional process improvement attempts based on total quality management principles

Figure 6.8. Actual group success rates in our study (i.e., employing e-collaboration) compared with those found by surveys of total quality management (TQM) and reengineering projects (conducted primarily face-to-face)

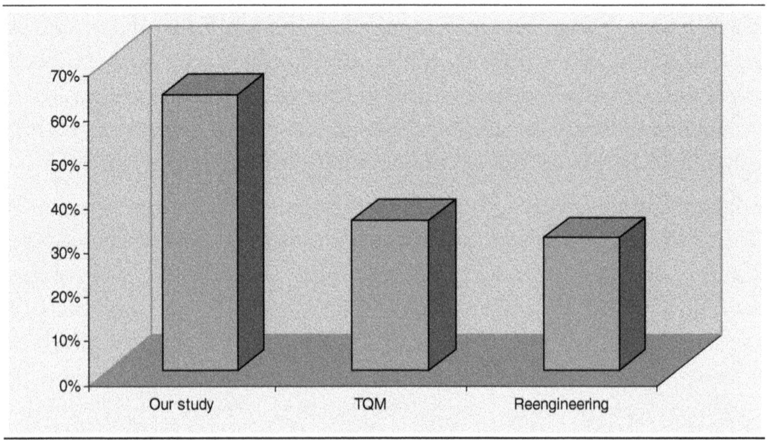

(Choi & Behling, 1997), and a large survey of traditional process improvement attempts by American and European companies employing business process reengineering principles (Champy, 1995). See Figure 6.8 for the results of this comparison, where success rates are shown side-by-side.

Some of the business process improvement groups we studied failed, at least according to our strict application of the criteria for success proposed in the process improvement literature (Burke & Peppard, 1995; Davenport, 1993; Hammer & Champy, 1993). Yet, Figure 6.8 suggests that the success rates obtained in our study (i.e., with e-collaboration technology support) are relatively high when compared to the success rates reported in the literature on total quality management and reengineering that I reviewed, which refers to business process improvement efforts conducted primarily by face-to-face groups. That literature reports success rates of business process improvement attempts based on total quality management principles ranging from approximately 20-34%. As for business process improvement attempts employing reengineering principles, the success rate obtained from the relevant literature was of approximately 30%.

That is, the rates of success of business process improvement attempts based on total quality management and business process reengineering principles have been consistently found to be around 34% or less, which is substantially less than the 62% rate (approximate) obtained in connection with our study.

This allows us to conclude that e-collaboration technology support seems to contribute to increasing the likelihood of success of business process improvement groups, which is overall fairly consistent with the perception-based evidence obtained from business process improvement group members, and discussed earlier in this chapter.

Summary and Concluding Remarks

The ever-growing amount of information flowing in organizations, knowledge specialization, and geographical distribution of expertise are among the main factors driving the expansion of the use of computers to support team-based business processes. At the same time, these factors also lead to more and more business processes being carried out by distributed teams. This scenario poses obstacles to traditional face-to-face interaction, which, combined with growing competitive forces, drive firms to increasingly use e-collaboration technologies to support business process improvement activities and knowledge sharing.

This chapter presents and discusses solid evidence that points to an increase of business process improvement group efficiency due to e-collaboration technology support. Efficiency gains are reflected in reduced group cost, lifetime, and reliance on managers. Additionally, the evidence presented and discussed in this chapter suggests that the number of simultaneous business process improvement groups that can be conducted in an organization is increased by e-collaboration technology support.

The evidence also points to slight (and, in some cases, not very significant) increases in perceived group outcome quality, and significant increases in perceived knowledge sharing effectiveness, as a result of e-collaboration technology support. These findings, combined with those regarding group efficiency, provide a sound basis for the endorsement of asynchronous e-collaboration technologies as tools for business process improvement and knowledge sharing.

Yet, there is also evidence that the communication medium provided by e-collaboration technology also poses obstacles for communication in business process improvement groups, which is difficult to reconcile with the generally positive outlook suggested by most of the evidence, particularly the evidence in connection with members' perceptions. This important issue is addressed later in this book in my discussion of what I like to call the "e-collaboration paradox."

Nevertheless, a possible alternative to address this issue is to break down complex tasks, such as business process improvement, into subtasks and use different e-collaboration media to support those subtasks. Prior research has shown that the amount of knowledge transfer involved in a communication interaction correlates the perceived difficulty in interacting through non-face-to-face media (Kock, 1998). Therefore, subtasks could be classified according to the amount of knowledge transfer involved, and assigned different e-collaboration media, where the degree of similarity of each medium to the face-to-face medium should be matched with the amount of knowledge transfer needed for effective completion of each subtask. That is, for high knowledge transfer subtasks, e-collaboration media that are very face-to-face-like (e.g., video-conferencing) should be used; whereas for low knowledge transfer subtasks, e-collaboration media that incorporate few of the elements found in face-to-face communication (e.g., e-mail) could be used.

Endnotes

1 The correlation coefficient found was .86, using the Pearson product-moment method, which suggests that more than 70% (the correlation coefficient squared) of the variation in the variable "information exchanges" can be explained through functional diversification. The likelihood that such high coefficient of correlation is due to chance was found to be lower than 5% (Kock et al., 1997).

2 This correlation has been found to be as high as .75 (Pearson). The likelihood that this high correlation is due to chance was found to be lower than 5% (Kock et al., 1997).

3 As with the rest of this book, all names and situations in this chapter have been disguised to honor confidentiality agreements.

4 The likelihood that the perception trend observed is due to chance is lower than 0.1%, as indicated by a statistical test called Chi-squared test.

5 The likelihood that the perception trend observed is due to chance is lower than 0.1%, as indicated by a Chi-squared test.

6 The theory actually claims that "uncertainty" and "equivocality" reduction are better accomplished through richer communication media. It is my interpretation that uncertainty is reduced through the exchange of information, and equivocality through the exchange of knowledge, according to the definitions of information and knowledge provided earlier in this book. This is the basis for my statement regarding media richness in connection with the communication of knowledge and information.

Chapter VII

The E-Collaboration Paradox

Paradoxical Results

This chapter advances a new explanation for some of the apparently contradictory findings discussed earlier in the book, which, in turn, reflect fairly well the body of mixed findings associated with academic and industry research on e-collaboration during the last 30 years (that research was usually conducted under other banners, such as computer-supported collaborative work and group support systems).

Those mixed findings suggest, although seemingly contradictorily, that: (a) people seem to consistently perceive face-to-face communication as posing fewer obstacles to effective communication than other, particularly electronic, media; and (b) when groups conduct collaborative tasks using e-collaboration technologies, they often present the same level (or better levels) of performance as (than) groups accomplishing the same tasks face-to-face—which, let's face it, is paradoxical in light of (a).

I argue here that the evolutionary history of humans suggests strongly that modern humans must have been largely "hardwired" for face-to-face communication, which, in turn, explains our perceptions favoring the face-to-face medium. Nevertheless, I also argue that human beings, when faced with communication obstacles, invariably try to compensate for them by changing their behavior in relatively predictable ways, which often leads to no negative impact on task outcome quality.

Later in this chapter, I will illustrate the explanatory power of these hypotheses based on evidence from an empirical study of business process redesign pairs, or groups of two individuals (sometimes referred to as dyads). The empirical study provides evidence that not only strongly supports the two hypotheses, but also cannot easily be explained otherwise, without resorting to the hypotheses and the underlying human evolutionary ideas that form the basis for the hypotheses.

Research on E-Collaboration

Particularly since the 1980s, e-collaboration has become an important topic for industry practitioners, and consequently, interest in the topic has been renewed among researchers. Several related fields of research, linked to particular groups and called different names, have addressed e-collaboration issues in the last 30 years. As mentioned before in this book, but which is worth emphasizing, the term e-collaboration is being used here as an umbrella term that comprises several other closely related fields, commonly known as computer-mediated communication, computer-supported cooperative work, groupware, group support systems, collaboration technologies, and, more recently, the so-called field of knowledge management.

The mid and late 1970s saw the development of the first personal computers (PCs), which were later connected to each other through the development of what came to be known as local area networks (LANs). Novell Corporation and later Microsoft played a major role in the shift from terminals connected to mainframes to PCs connected through LANs with the development of the first commercially available network operating systems. This led to the development of many synchronous and asynchronous e-collaboration technologies in the 1980s.

Some of these e-collaboration technologies, such as Information Lens and the Coordinator, extended the basic features present in early e-mail systems. Others provided support for decision-making meetings, such as GroupSystems, Teamfocus, and MeetingWorks. Still others, such as Lotus Notes and Domino, were suites on which customized e-collaboration systems could be developed to support group processes. Much industry-oriented and invention-based research was conducted during this period in universities, government agencies, and corporate research centers. This research led to most of the e-collaboration technologies developed during this period.

The 1970s also saw the birth of more systematic academic research on the impact of e-collaboration tools on people. Among the pioneers were Murray Turoff (1973, 1975, 1978) and Starr Roxanne Hiltz (1978). They conducted their early investigations based on the electronic information exchange system (EIES), one of the first e-collaboration systems to be developed and the precursor of modern Web-based asynchronous conferencing systems (Turoff, 1978). Increasing interest in the impact of e-collaboration technologies on people and organizations followed their initial investigations. Soon after, Paul Cashman and Irene Greif organized one of the first academic conferences on e-collaboration in 1984, called the computer-supported cooperative work (CSCW) conference, which has since become one of the key outlets for e-collaboration research focusing on human and technological issues (Bannon, 1993; Grudin, 1994).

A separate e-collaboration research tradition, focusing on technological and behavioral issues in connection with group decision support systems (GDSS), began in the late 1970s and early 1980s. By way of contrast with CSCW research, whose main motivation was to solve technological obstacles to e-collaboration and analyze the effects of those solutions on people, GDSS research emerged from the work of information systems researchers in business schools. Among the pioneers of GDSS research are Jay Nunamaker and Gerardine DeSanctis. Research on GDSS went on to become one of the most important areas of research in the incipient field of information systems. At the same time, CSCW research expanded its base, while still retaining its technological focus, leading to a chasm between GDSS and CSCW research that remains to this day.

With the emergence of the Internet in the early 1990s, research and development related to e-collaboration technologies extended beyond traditional research organizations and became a central business issue. As mentioned before in this book, the Internet began its exponential growth in 1993, when one of the first Web browsers, called Mosaic, was developed. This set the stage for the migration of e-collaboration systems from LANs to the Internet, mostly as client-server systems running on platforms made up of generic, platform-independent Web browsers (on the client side), and platform-dependent Web servers (on the server side).

Today, the Web browser interface has become one of the most widely used interfaces for e-collaboration technologies. The advent of the Internet led to a technological revolution that seems to have caught both CSCW and GDSS researchers largely by surprise. By the mid-1990s, many powerful and affordable e-collaboration tools and platforms became commercially available all over the world, which opened the door for a wide range of researchers from many countries to add to the body of e-collaboration research. Many of these researchers came from disciplines where research on e-collaboration was uncommon, such as accounting, medicine, organizational psychology, and marketing, to name only a few.

The E-Collaboration Paradox

The research summarized here has led to many empirical findings. Two general and competing findings, in particular, have led to much speculation and have characterized what is referred to here as the "e-collaboration paradox."

(a) People in general seem to consistently perceive face-to-face communication (as well as communication that incorporates key elements of the face-to-face medium, such as the ability to use non-verbal cues to convey ideas) to pose fewer obstacles to effective communication than other, particularly electronic, media.

(b) When groups conduct collaborative tasks using e-collaboration technologies, they often present the same level of performance or even perform better than groups accomplishing the same tasks face-to-face, which is contradictory with notion (a). This, of course, also is supported strongly by the evidence in connection with the effects of e-collaboration technologies on business process improvement groups, discussed earlier in this book.

While the e-collaboration paradox has led to much speculation, it can be argued that the paradox can easily be explained based on two hypotheses, which are consistent with the competing findings already summarized.

The first hypothesis, referred to here as the media naturalness hypothesis, argues that the human communication apparatus has been designed primarily for face-to-face communication, and that, because of such design orientation (which has evolutionary reasons), tools that suppress too many of the elements present in face-to-face interaction end up posing cognitive obstacles for communication.

The second hypothesis, called here the compensatory adaptation hypothesis, argues that human beings invariably adapt their communicative behavior in order to overcome obstacles posed by e-collaboration technologies, often leading to an interesting result—they perform just as well or even better than they would face-to-face, even though their perceptions of the e-collaboration tools used still match predictions based on the media naturalness hypothesis. That is, they see e-collaboration tools as generally posing cognitive obstacles to communication.

Media Naturalness: Human Beings Have Not Been Designed for E-Collaboration

The core of the theoretical argument presented here is relatively simple. I am essentially arguing that e-collaboration technologies, in general, offer certain advantages, such as allowing for asynchronous and distributed group interaction, and, at the same time, pose obstacles for communication in groups. If there were no advantages, it would be difficult to explain the widespread and growing use of e-collaboration technologies today. The obstacles, it is argued here, are due to the fact that our biological communication apparatus, which includes specialized organs and brain functions, has been optimized by Darwinian evolution for face-to-face communication incorporating five main elements—co-location, synchronicity, and the ability to convey body language, facial expressions, and speech.

Evidence about the evolution of our biological communication apparatus suggests that during over 99% of our evolutionary cycle, our ancestors relied on co-located and synchronous forms of communication through facial expressions, body language, and sounds (including speech, which uses a large variety of sound combinations) to exchange information and knowledge among themselves (Boaz & Almquist, 1997; Cartwright, 2000). According to evolutionary principles, a plausible conclusion from this is that our biological communication apparatus was designed primarily to excel in face-to-face communication.

The above conclusion is supported by the presence of obvious face-to-face communication adaptations in our biological communication apparatus. For instance, evolution endowed human beings with a complex web of facial muscles that allows them to generate over 6,000 communicative expressions. It is reasonable to assume that this complex muscular structure is designed to help human beings to communicate with each other, since very few of those muscles are used for purposes other than communication, such as chewing (Bates & Cleese, 2001; McNeill, 1998).

The evolutionary path that led to our species suggests a noticeable evolutionary direction over millions of years toward the development of a biological communication apparatus that supported ever more sophisticated forms of speech, culminating with the development of complex speech approximately 100,000 years ago. The advent of complex speech was enabled by the development of a larynx located relatively low in the neck and an enlarged vocal tract (key morphological traits that differentiate modern humans from their early ancestors and that allow modern humans to generate the large variety of sounds required to speak most modern languages) (Laitman, 1984, 1993; Lieberman, 1998). The morphology of the human ear also suggests a specialized design to decode speech (Lieberman, 1998; Pinker, 1994).

The evolution of closely matched brain and body functions, which follows from the widely held brain-body co-evolution law of modern evolution theory (Lieberman, 1998; Wills, 1989, 1993), provides a scientific basis for the apparent bias toward face-to-face communication, and is reflected in the media naturalness hypothesis.

Individuals who choose to use e-collaboration tools to accomplish collaborative tasks experience increased cognitive effort and communication ambiguity proportionally to the degree to which the tools suppress elements that are present in face-to-face communication (e.g., synchronicity, ability to convey/perceive non-verbal communication cues).

Even though this hypothesis is task independent (i.e., it applies to all collaborative tasks), it acknowledges that the link is less noticeable in tasks that do not involve intense communication, which are seen as tasks that involve little knowledge sharing among collaborators (Kock, 2001, 2001b).

The media naturalness hypothesis links the use of e-collaboration tools with high cognitive effort and communication ambiguity, but not necessarily with specific media choices (e.g., avoidance of certain media) or task-related outcomes (e.g., successful business process redesigns). In doing so, it explains empirical findings that support, in part, two influential organizational communication theories, namely the social presence and media richness theories (Daft & Lengel, 1986; Daft et al., 1987; Short et al., 1976), while at the same time avoiding the problems associated with making predictions about media choice or task-related outcomes based on communication media traits, which led to much criticism against the social presence and media richness theories by social researchers (El-Shinnawy & Markus, 1998; Lee, 1994; Markus, 1994). For example, the media naturalness principle is compatible with the notion that social influences can lead users to modify their behavior (Lee, 1994; Markus, 1994; Ngwenyama & Lee, 1997) in ways that are independent of the apparent degree of naturalness of a medium, even though the cognitive effort required for this to happen will be higher than if a more "natural" communication medium (i.e., the face-to-face medium) were used.

Compensatory Adaptation: Human Beings Often Try to Compensate for Obstacles Posed to Them

It is intuitive to think that obstacles to high communication effectiveness in collaborative tasks lead to lower quality of task outcomes. However, there is a wealth of evidence from fields as diverse as biological anthropology (Dobzhansky, 1971) and analytical psychology (Jung, 1968) suggesting that human beings voluntarily and involuntarily compensate for obstacles posed to them, in many cases overcompensating for those obstacles and achieving even better outcomes than if the obstacles were not present. This phenomenon has the potential to contradict deterministic predictions linking negative communication media influences on group effectiveness with low group outcome quality.

Kock (1998, 2001b) obtained empirical evidence of this compensation phenomenon in connection with e-collaboration, some of which has been summarized previously in this book. The evidence was obtained in the context of a study that compared groups performing complex and knowledge-intensive tasks (e.g., business process improvement groups) over an e-mail-based e-collaboration tool and face-to-face. As discussed before in this book, the medium created by the e-collaboration tool was consistently seen by group members as less appropriate than the face-to-face medium to accomplish the tasks. Yet, the tasks accomplished through the e-collaboration medium yielded outcomes that were perceived to be of equal or better quality than those produced by the face-to-face groups.

The compensatory adaptation hypothesis argues that users of e-collaboration tools present two common patterns of reaction toward those tools. First, the users in general perceive those tools as creating communication media that pose cognitive obstacles to communication when compared with the face-to-face medium (Kock, 2001, 2001c), as proposed by the media naturalness hypothesis discussed in the previous section. That is, even though electronic communication tools may reduce or eliminate physical obstacles to face-to-face communication (e.g., e-mail and instant messaging, which allow people to communicate instantly over long distances), they also increase the cognitive effort required to communicate information and knowledge. The second common pattern of reaction is one of compensation for the obstacles posed by the media (Kock, 1998, 2001b), which is embodied in the compensatory adaptation hypothesis.

Individuals who choose to use e-collaboration tools to accomplish collaborative tasks tend to compensate for the cognitive obstacles they perceive as associated with the lack of naturalness of those tools, which leads them to generate group outcomes of the same or better quality than those generated through the face-to-face medium.

In conclusion, the media naturalness and compensatory adaptation hypotheses argue, in an apparently paradoxical way, that obstacles posed by e-collaboration tools do not significantly decrease the quality of group outcomes, as groups (often involuntarily) attempt to compensate for them, or even to overcompensate for them. This counterintuitive notion explains the contradictory empirical findings of research on e-collaboration technologies conducted so far, and arguably opens the way for future theoretical progress and integration of seemingly disparate theories of e-collaboration behavior.

Evidence from an Empirical Study of Business Process Redesign Pairs

When I first proposed some of the ideas above in an article that was aptly (in my view) titled "The Ape that Used Email" (Kock, 2001c), there was some skepticism from the academic community, which I didn't take personally at all. There is no doubt in my mind that the hypotheses presented here are quite new and somewhat controversial, and thus, some skepticism toward them can be expected. Since there are not many studies that directly address both hypotheses in the same context, the discussion of some preliminary evidence supporting them is warranted. This is what I try to accomplish in this section, by discussing a targeted study, which adds to the body of evidence presented in previous chapters.

The hypotheses presented in this chapter were tested through a field experiment employing a repeated measures design where the communication medium used varied according to two experimental conditions: face-to-face and electronic. I essentially asked individuals to perform similar tasks using the two different media, which is what I mean by the term "repeat measures design." The impact of changes in the communication medium factor on a set of dependent variables was assessed by means of multiple one-way analysis of variance (ANOVA) and Mann-Whitney U tests (Green et al., 1997; Rosenthal & Rosnow, 1991), which are statistical tests that are often grouped under the general designation of comparison of means tests. These are statistical analysis tests of the type used by pharmaceutical companies to compare the effects of the administration of a new drug to a group of patients with those of the administration of a placebo to the same group (at a different time) or to a different group of patients (also called the "control group").

The research study involved subjects with substantial hands-on experience in a particular type of collaborative task (namely, business process redesign), where individuals analyze and redesign a business process. This is the task that precedes the new business process implementation in business process improvement efforts. The subjects were recruited from management and engineering ranks of a large defense contractor located in the northeastern US. The subjects were familiar with each other and with the electronic communication medium used prior to their participation in this field experiment. However, they had no prior experience using the electronic communication medium for the collaborative completion of tasks of the same type as, or even similar complexity to, the experimental task. The subjects' ages ranged from 23 to 60, with a mean age of 35. Fifty-nine percent of the subjects were males.

The subjects were randomly assigned to dyads (or groups of two individuals) and to communication media conditions. Each dyad completed two similar business process redesign-related tasks using different communication media for each task. Half of the dyads (i.e., 10 dyads) completed one of the tasks face-to-face, while the other half completed the same task electronically. After this, all dyads moved on to the next task, using different media than they had used in the previous task (i.e., the dyads previously interacting face-to-face now interacted electronically and vice-versa).

Figure 7.1 summarizes the results of the analysis in the form of averages (or means), all on a scale from 1 to 7 (some variables had to be transformed to fit this scale), in connection with five key variables: cognitive effort, communication ambiguity, message preparation, task outcome quality, and fluency. The variable "message preparation" refers to the perceived level of carefulness with which electronic messages or oral contributions are prepared. The "fluency" variable refers to the number of words conveyed through a communication medium per unit of time. The other variables are self-explanatory and used analogously to other similar variables discussed earlier in this book. Consistent with Kock (1998), message preparation and fluency were used as indicators of compensatory adaptation. Task outcome quality was validated through inter-coder reliability; that is, more than one person was asked to assess the task outcomes of each group, and the scores generated by different people were correlated. The other constructs had their reliability assessed through the test-retest method, where questionnaires are administered at different points in time, and their answers are correlated (Rosenthal & Rosnow, 1991).

All of the measures in Figure 7.1 have been analyzed based on comparison of means tests, which compare evidence obtained from different samples and

Figure 7.1. E-collaboration technology effects on several variables

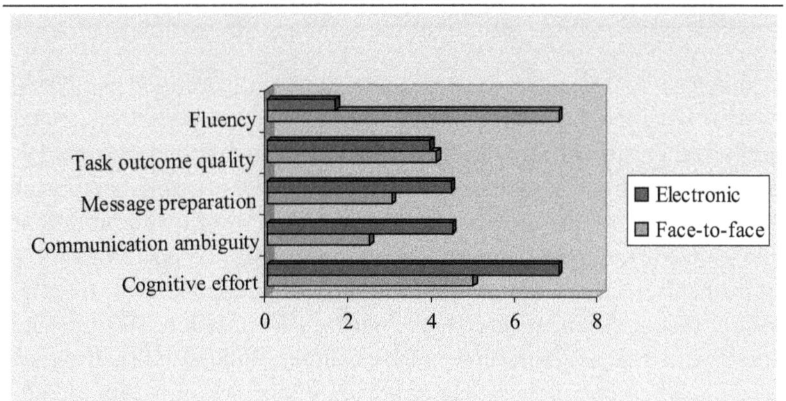

collected under different conditions. The two comparison-of-means tests used were, as mentioned before, one-way ANOVA and Mann-Whitney U tests. The differences between means were found to be statistically significant for all constructs except task outcome quality.

The study results suggest that the use of an e-collaboration tool, when compared with the face-to-face medium, increased cognitive effort by about 41%, communication ambiguity by about 80%, and message preparation by about 47%, while at the same time reducing fluency by approximately 77%. The study also suggests that the use of the e-collaboration tool had no significant impact on the quality of the outcomes generated by the dyads.

These results are consistent with the media naturalness and compensatory adaptation hypotheses. The differences in cognitive effort and communication ambiguity provide strong and direct support of the media naturalness hypothesis. The differences in message preparation and fluency strongly suggest that the subjects thought more about what they were saying when using the e-collaboration tool than when they interacted face-to-face (i.e., more carefully prepared their communication messages), and thus provide general support for the compensatory adaptation hypothesis. That is, it is reasonable to expect that the extra preparation was an attempt to compensate for the obstacles posed by the electronic medium. Finally, the fact that the differences in task outcome quality (i.e., the quality of the process redesigns produced by the dyads) were insignificant, suggests that the compensatory adaptation behavior displayed by the subjects reached its goal.

Summary and Concluding Remarks

It is important for managers to understand the "e-collaboration paradox" and how it can be explained by the media naturalness and compensatory adaptation hypotheses. The reason is that empirical evidence such as that discussed in the previous section may easily lead managers to believe that e-collaboration tools, perhaps because of the tools' effectiveness in supporting communication interactions, do not affect group outcomes, neither directly nor indirectly. This will probably lead to frustration when those managers realize that "good [electronic] communication requires hard work" (Bartlett, 2001, p. 1). The evidence and discussion presented in this chapter, coupled with that presented in previous chapters, suggest that the apparently neutral impact of e-collaboration technology support on the outcomes of collaborative tasks is the result of a complex interplay of e-collaboration technology effects on mediating variables.

Nevertheless, one implication for managers of the discussion presented in this chapter, as well as similar discussions in previous chapters, is that small groups conducting complex and knowledge-intensive tasks such as business process redesign can be conducted entirely electronically and still be successful. Given the emergence of the Internet and the consequent multiplication of organizational forms characterized by their low dependence on physical structures for employee interaction, such as the so-called virtual organizations, this is not only good news for organizations, but also provides the basis on which to call for increasing use of e-collaboration tools to support all types of group tasks, ranging from routine group tasks, where the use of e-collaboration tools is already relatively common, to more ad-hoc (or project-based) ones, where the use of e-collaboration tools is still rare.

Moreover, the discussion in this chapter supports the idea, presented earlier in this book, that e-collaboration tools are likely to be beneficial to business process improvement groups. However, an important issue that needs to be addressed in future research is the likely reduction in cognitive effort associated with e-collaboration tools as users become more adept at using those tools.

Compensatory adaptation has a cognitive cost. As past research findings on human cognition suggest, learning associated with tasks that impose cognitive costs usually reduces future cognitive costs associated with the tasks in question; that is, it is likely that repeated use of an e-collaboration tool will invariably reduce the need for compensatory adaptation in connection with that tool in the future. At least one study addressed this issue (Carlson & Zmud, 1999). The study supports the prediction that the need for compensatory adaptation will decrease as the familiarity with an e-collaboration tool increases, but does not provide much insight on how strong this moderating influence is relative to the media naturalness influence.

I hope that it will be clear from this chapter that, in spite of the large amount of research on e-collaboration conducted so far, this field of research is still full of vitality. Many theoretical issues remain to be resolved, which in my view cannot be accomplished without much additional empirical research. E-collaboration is likely to continue to be a fertile area of research in the foreseeable future. What is needed is creativity to identify and address core questions that still remain to be answered.

Chapter VIII

Successful Business Process Improvement through E-Collaboration

The Ubiquity of Business Process Improvement

The idea of business process-focused improvement has been with us for many years. Many speculate that it is as old as the total quality management movement, which began in Japan in the 1950s. Some think the idea is much older, dating back to the time of the Pharaohs of Egypt.

At least two popular management movements incorporated the business process improvement idea. One of these management movements is the previously mentioned total quality management, whose popularity reached a peak in the U.S. in the 1980s, and whose main figure was William Deming. The other is business process reengineering, which was developed and became very popular in 1990s, and whose main proponent has been Michael Hammer.

Examples of business process-focused improvement have been and still are found in abundance in the organizational world, whether the organizations considered are public or private, for profit or not. Therefore, there is plenty of organizational data that can be used to establish how successful business process-focused improvement projects have been so far. Moreover, as the total quality management reached its peak of popularity in the 1980s, there is now long-term organizational evidence that allows for the assessment of the outcomes of total quality management projects over several years.

Choi and Behling (1997) have summarized studies of total quality management. This summary suggests some lackluster results. For example, a survey of 500 manufacturing and service firms in the US indicates that two-thirds of their executives believe that total quality management programs have not made them more competitive. Another survey of 100 British firms, where total quality management programs have been implemented, indicated that only one-fifth perceived their programs as significantly impacting their firms' performance. Yet another survey conducted by the American Electronics Association suggests that the popularity of total quality management programs among its member firms dropped from 86% in 1988 to only 73% in 1991. The same survey revealed that in 63% of the firms, total quality management programs that had been in operation for an average of 2.5 years had reduced defects by no more than 10%. An evaluation of total quality management programs by McKinsey & Company found that two-thirds of them had been discontinued due to their results falling short of the original expectations.

On the other hand, Hendricks and Singhal (1997) conducted a study where quality award-winning companies were compared to a sample of control firms. Their study spanned a period of 10 years for each company, comprising six years before and three years after the quality awards were conferred. One of their findings was that the mean change in the operating income (i.e., sales revenues minus production, depreciation, sales, and administrative costs in connection with the products sold) for the award-winning firms was 107% higher than that of the control sample. Another finding was that the mean increase in sales had been 64% higher for the award-winning firms, which suggests that firms that have won quality awards have done considerably better on sales growth than the control firms. On the other hand, only weak evidence was found that the award-winning firms had been more successful in controlling costs than the firms in the control sample.

The business process reengineering history is somewhat different. It started with case studies showing the almost miraculous potential of the combination of a focus on business processes and a radical approach to change (Hammer, 1990; Hammer & Champy, 1993). Yet, not too long after these initial case studies were published, a survey of private firms in the US and Europe, conducted at the peak of the reengineering movement by one of its forerunners (Champy, 1995), indicated a failure rate as high as 70%.

No doubt, mixed findings exist regarding the success and failure of business process improvement approaches and projects, whether they resemble total quality management, reengineering, or any other less known business process-focused improvement approach. Yet, business process improvement is still so widely practiced among organizations of all sizes that one could wonder whether its bad side is being ignored or if there is nothing better around. A topic closely related to total quality management, ISO 9000 certification, is as popular as ever, particularly among exports-oriented companies and satellite part suppliers of large manufacturers (Kock & McQueen, 1997; Ruddell & Stevens, 1998).

Information technology companies that built and marketed products around the business process reengineering idea (i.e., Germany's software giant SAP), in the period between the start of the reengineering movement and 2000, have done as well as any company in their industry from a financial perspective—their success was nothing short of astounding. For example, SAP reported a record 1.67 billion Deutsche Marks (929.9 million US dollars) in pretax profit in 1997, compared with 967 million Deutsche Marks in 1996. SAP's sales in 1997 were 6.02 billion Deutsche Marks, up 62% and also a record (Rose, 1998).

My conclusion is that although different measures of success will yield different and often conflicting results, the trend toward business process improvement will remain strong for at least one or two more decades. After the barrage of criticism of business process-based improvement methods by business specialists and company strategy commentators in the mid-1990s, many management experts now seem to acknowledge this inevitable trend. Moreover, many information technology experts working in hot areas also seem to hold similar views, which they air as often as they can. A good illustration of such statements is Umar's (1997), in the introductory chapter of his book on client/server environments. Umar summarizes the organizational demands and circumstances surrounding and sustaining the trend towards business process-focused improvement as follows:

Enterprises in the 1990s and beyond are typically characterized by flatter organizational structures, increased demands for flexibility, pressures to respond quickly to market conditions, intense local and global competition, and continued business process reengineering and improvements for enterprise efficiency.

(Umar, 1997, p. 4)

Among other things, new developments in information technology have been both supporting and feeding the growing importance of business process improvement in organizational circles. Information technology is not only a tool that enables organizations to deal with change by supporting the implementation of new business processes, it also creates competitive disparities among organizations that feed the pressure for change and, as a result, the need for business process improvement.

There are some technology and business pundits who seem to disagree with this view, arguing that information technology is becoming a commodity that truly cannot be used to improve an organization's competitive position (Carr, 2003, 2003a). Those pundits, however, are still a very small minority.

Finally, we cannot forget that business process improvement can potentially build on a well-established foundation of techniques for systems analysis and design, whose popularity increases as the use of information technology in organizations grows. Why is there such an increase in popularity? One of the reasons is that those techniques are fundamental for the success of new business process implementation and related automation projects. I do not see any signs of a possible reduction in the use of computers and computer applications in organizations in the near future. On the contrary, an accelerated growth is the most likely prospect.

However, traditional systems analysis and design approaches need to be adapted to serve as a solid methodological basis for business process improvement efforts. The problem is that traditional systems analysis and design approaches have been developed to support the automation of existing business processes in organizations, without any concern for business process redesign. Those methodological development efforts were based on one mistaken idea, which seemed to be responsible in part for the software crisis of the 1980s and 1990s that led in part to the emergence of the business process reengineering movement. The idea was that the simple automation of manual business processes usually yields significant gains in business process quality and productivity.

The modification of traditional systems analysis and design techniques to support business process improvement efforts, especially modification efforts that take into consideration the increasingly information-intensive nature of modern business process, are particularly important in reversing this trend. There have been small but steady steps in that direction (Danesh et al., 2003; Kock, 2003; Kock & Danesh, 2003), and I hope that the pace of those developments will increase in the future.

The Trend Toward Organizational Learning

One of the key ideas behind the organizational learning movement (Argyris, 1992; Senge, 1990; Senge et al., 1994) is that organizations (as collective entities) acquire, store, and use knowledge over time. It is now widely accepted that an organization's competitive advantage depends heavily on its ability to acquire, store, and use unique knowledge in a way that allows the organization to produce and deliver products (e.g., goods, services, information) that have a singular appeal to its customers (Spender, 1996; Prahalad & Hamel, 1990).

Has not this notion been commonplace for a long time? Have not many people known for years that knowledge management and organizational competitiveness are closely related? I believe the answer is yes. Nevertheless, we have seen since the mid-1990s a widespread and growing interest in organizational learning, with an intensity that has not bee seen before. Why is that so? Is it because organizational learning is a new fad? Certainly not, as the concept has been proposed and discussed in the business literature, as well as offered as a management consulting product, at least since the 1970s. As mentioned previously in this book, one of the original proponents of the organizational learning approach is Chris Argyris, a Professor Emeritus at Harvard University.

The key to explaining the present interest in organizational learning lies, in my opinion, in two factors. One, which is eminently socioeconomic, is the accelerated growth of the body of knowledge that is relevant to the creation of goods and services, as well as other organizational outputs such as software and information. The other factor is closely related to how the human species evolved, from a biological perspective—namely, the current human cognitive limitations. Let me explain the nature of these factors and how they affect our interest in management approaches such as organizational learning.

Speculations about the speed with which the available body of knowledge grows vary widely. Some believe that it doubles every two years, while others speculate that it triples every 18 months. Measures of knowledge also vary widely, from the number of registered patents to the number of articles published in academic journals. Given this, and until we have a widely accepted knowledge measurement unit, it is meaningless to try to estimate precisely how much knowledge is created every year, or the rate at which knowledge is created. Yet, at least two things are obvious.

First, the amount of knowledge created in any generic field of research every two years is very large; certainly larger than what an average person would be able to absorb without some form of professional specialization (i.e., a focus on specific knowledge within one or a few fields). For example, the amount of knowledge in the field of distributed computing, which practically did not exist in the 1950s, had grown so large by the 1990s that it spun off a number of subfields, such as that of client-server computing.

Second, the speed with which knowledge is created has been increasing. This is a natural outcome of the number of people with higher education degrees worldwide. As some of those people make new discoveries in their fields, that, in turn, adds to the body of knowledge in those fields. In other words, much less knowledge, however it is measured, was created per year in the 1950s than in the 1990s.

I already pointed out that people specialize in subfields, often creating a number of them in the process, because they cannot assimilate the growing body of knowledge being created at higher-level fields. That is, specialization is a function of knowledge growth and cognitive limitations. And specialization leads to knowledge fragmentation or, in other words, the creation of new subfields of knowledge.

Where does the growing interest in organizational learning fit in this picture? Knowledge fragmentation leads to functional fragmentation in organizations. For example, John is the new Web site administrator, an organizational function (or job description) that entails a well-defined set of responsibilities that, in turn, requires specialized knowledge. This organizational function did not exist in the 1980s, because the Web became popular in the 1990s.

The problem is that John must hold a considerable body of knowledge and keep up with the new knowledge being created by the day (if not by the hour) in his area, the area of Web site administration. Hence, John cannot combine this organizational function with other related functions, such as the more established function of network administrator. Network administration then has to be done by another person (let's say Karen).

This is a clear example of the type of functional fragmentation that occurred late in the 1990s, and that still keeps occurring to this day, apparently at an ever-increasing pace. The old role of network administrator has now spawned at least one related role, namely that of Web site administrator, due to the creation of new knowledge about Web site technology. The problem is that John and Karen are still going to be involved in a few common business processes, like providing computer support to internal customers who need network access rights to set up their own Web pages. As such, John and Karen need to share some of their specialized knowledge. Here is where organizational learning comes in, as interfunctional knowledge sharing is one of the most important aspects of organizational learning (Kock & Davison, 2003; Redding & Catalanello, 1994).

The Emergence of Virtual Organizations

As discussed, there are noticeable organization trends toward employing business process improvement and organizational learning for improving competitive advantage, and these are trends that seem unlikely to fade away soon. This is because both business process improvement and organizational learning allow organizations to face problems that are typical of highly competitive and knowledge-intensive economies. In the years to come, the terms used for business process improvement and organizational learning may change (e.g., some have tried in the late 1990s and early 2000s to subsume organizational learning into a newer field called knowledge management), but the focus on business process-based change and knowledge acquisition and deployment will continue.

However, a third major trend, discussed previously in this book, must be brought into the picture—a technological trend. Since the mid-1980s, there has been an accelerated increase in the use of computer networks to provide communication links within and between organizations. Electronic communication links within organizations have been established through local area networks (LANs) and, more recently, Web-based intranets. Similar links between organizations have been established through wide area networks (WANs) and, more recently, the Internet, the Web, and interorganizational computer systems all over the world (e.g., the worldwide system that enables money transfers between banks based on standardized routing numbers).

As discussed earlier in this book, an accelerated increase in the number of knowledge specializations and their globalization has accompanied this worldwide diffusion of computer networks, which has made it possible for a number of organizations to become independent of geographical constraints. This, in turn, has led a number of organizations to structure themselves around communication networks. Such organizations also moved towards organizational paradigms that place emphasis on flexibility, knowledge accumulation and deployment, and distributed teamwork. One such paradigm, and perhaps the best known in this category, is the virtual organization paradigm (Davidow & Malone, 1992; Kock, 2000a, 2002; Mowshowitz, 1997).

The concept of virtual organizations has become one of the hottest management topics of the 1990s and early 2000s, particularly given the possibilities afforded by local and wide area networks of computers. Advantages of moving from physically aggregated organizational units toward electronically linked ones have been widely publicized. Facilities rental and maintenance costs can be considerably reduced or eliminated. Employees can work from their own houses or from distributed offices near their houses. Company offices can spread over large geographical areas, reaching a larger number of customers than physically aggregated offices would. Operations data, while being accessed in a decentralized fashion, can still be stored and managed in a centralized manner. Clients can purchase goods, services, and information irrespective of where they are in the world, without the need to travel long distances.

Given the three trends discussed so far in this chapter (i.e., trends toward business process improvement, organizational learning, and virtual organizations), it is reasonable to expect that there will be increasing pressure on organizations to find ways to carry out business process improvement and knowledge sharing (a key component of organizational learning) in a distributed, e-collaboration technology-mediated fashion. The alternative to that will be to conduct thousands of face-to-face business process improvement and knowledge sharing meetings to face competitive pressures and the explosion of knowledge. Unless we find out how to beam people around (like in Star Trek) at a very low cost, this is unlikely to be a feasible alternative.

Previous chapters have discussed evidence that suggests that e-collaboration technology support is likely to have a positive effect on business process improvement and knowledge sharing. Yet, that evidence points to the fact that sometimes e-collaboration technology support does not prevent business process improvement groups from failing. Given that, perhaps it is time to try to understand the success factors associated with distributed and e-collaboration technology-supported business process improvement and knowledge sharing.

Success Factors: An Analysis of Twelve Groups

In previous chapters of this book, I discussed the link between business process improvement and knowledge sharing. I also discussed the impact of e-collaboration technologies on both business process improvement and knowledge sharing. Now I will complement those discussions with an investigation of what makes distributed business process improvement groups succeed with the support of e-collaboration technologies. As in previous chapters, business process improvement groups conducted at two New Zealand organizations, Waikato University and MAF Quality Management, are analyzed here. As it has been shown before in this book, the communicative behavior of group members observed in those two New Zealand organizations was remarkably similar to the behavior observed in US organizations (and consistent with a theoretical framework proposed by me in this book), which provides the basis on which the findings in this chapter can be presented as somewhat general and likely to repeat themselves in different organizational environments. The reader, of course, should see this as a reasonable supposition, and not a statement that can be made with absolute certainty. The truth seems to be that each organization has its idiosyncrasies, which often prevent solutions that work in other organizations from having exactly the same outcomes.

As with much of the evidence presented in this book, the evidence discussed in this chapter is a refinement of previous analyses conducted by me, and whose preliminary results appeared in conference proceedings and journal articles. Such previous analyses helped me unveil patterns on which I focused my attention later on. In the case of success factors, previous analyses led me to pay particular attention to three types of factors:

- Leadership factors, which relate to characteristics of the leaders (or moderators) of e-collaboration technology-supported business process improvement groups.

- Membership factors, which relate to group membership attributes, such as the general level of group member interest in business process improvement.
- Other factors, which refer to general characteristics of each business process improvement group, including characteristics of the business process targeted by the group.

In my analyses of interviews conducted with group members and other individuals, most of whom interacted with group members, I tried to summarize perception patterns related to group success. This summarization led to the building of three main tables, where each factor was split into a few subfactors or variables. These variables were then assigned qualitative values along a three-point categorical scale: low, medium, and high. The results were matched against the content of the dependent variable "group success," which refers to the general level of success of each group, for each factor separately. The three following sections discuss the findings that emerged from the analysis.

Leadership Factors

Table 8.1 shows the variation of leader-related variables among the 12 groups, together with the degree of success obtained by each group. Groups are ordered on the table based on their degree of success, the most successful being at the top of the table, and the least successful at the bottom. From left to right, the three first columns of Table 8.1 show: a brief description of each business process[1] targeted for redesign, the organization to which each business process belongs, and the degree of success obtained by each group. The last three columns on Table 8.1 show the leader attributes that emerged from the interviews I conducted with business process improvement group members as the most important (or relevant) leader attributes in the context of e-collaboration technology-supported business process improvement. Those attributes were the leader's relative status among all group members, the leader's degree of attempted control of the group discussion, and the degree of involvement of the leader in the business process being targeted for redesign.

It can be inferred from Table 8.1 that, when taken individually, neither the leader's relative status or attempted control consistently affected group success. On the other hand, the business process involvement of the group leader was consistently related to a high degree of group success. With the exception

Table 8.1. Leadership factors impacting the success of e-collaboration technology-supported groups

		Leader's			
Process Description	**Organi-zation**	**Success**	**Relative Status**	**Attempted Control**	**Process Involvement**
Newsletter editing	MAF	High: Group agreed on process changes, from which all were implemented with positive results	High: Senior manager	Low: Asked and individually thanked members for opinions, let the discussion flow	High: Process owner
University course	Waikato	High: Group agreed on process changes, from which all were implemented with positive results	Low: Part-time graduate assistant	Low: Asked for opinions and let the discussion flow	High: Process owner
Student assignment handling	Waikato	High: Group agreed on process changes, from which all were implemented with positive results	Medium: Division manager	Low: Asked and individually thanked members for opinions, let the discussion flow	High: Process owner
International graduate student support	Waikato	High: Group agreed on process changes, from which all were partially or fully implemented with positive results	Medium: Division manager	High: Half-way through the discussion, unilaterally changed its topic	High: Process owner
International student adaptation support	Waikato	High: Group agreed on process changes, from which all were partially of fully implemented with positive results	Medium: Middle manager	Low: Asked and individually thanked members for opinions, let the discussion flow	High: Process owner
IT users support	MAF	High: Group agreed on process changes, from which three quarters were implemented with positive results	Medium: Senior computer support person	Medium: Tried to keep group discussion on focus	Medium: Key process member

Table 8.1. Leadership factors impacting the success of e-collaboration technology-supported groups (cont.)

		Leader's			
Process Description	**Organi-zation**	**Success**	**Relative Status**	**Attempted Control**	**Process Involvement**
Software support	MAF	High: Group agreed on process changes, from which two-thirds were implemented with positive results	Medium: Middle manager	Low: Asked for opinions and let the discussion flow	High: Process owner
Pest/disease outbreak communication	MAF	High: Group agreed on process changes, from which one-third were implemented with positive results	High: Senior manager	Low: Asked and individually thanked members for opinions, let the discussion flow	Medium: Key process member
Undergraduate academic support	Waikato	Medium: Group agreed on process changes, from which none was implemented	Medium: Senior liaison person	Low: Asked for opinions and let the discussion flow	Low: Marginally involved in the process
Quality management consulting	MAF	Low: No agreement on process changes was achieved	Low: Middle manager (amid several senior managers)	Low: Asked and individually thanked members for opinions, let the discussion flow	Medium: Provided process-related training
Staff training and development	MAF	Low: Unfocused discussion, the group was discontinued without agreeing on any process changes	High: Senior manager	High: Tried to obtain specific information from members based on individual assumptions	High: Process owner
Student computer support	Waikato	Low: No agreement on process changes was achieved	Low: Part-time graduate assistant	Medium: Pointed out digressions, tried to keep group discussion on focus	Low: Marginally involved in the process, formerly a member of process team

of the group whose target was the business process related to staff training and development at MAF Quality Management, leaders who owned (i.e., formally managed) the business processes targeted by their groups were able to lead their e-collaboration technology-supported groups to a successful completion. There were also two examples of business process improvement groups in which the leaders were key business process team members (i.e., members of the teams that executed the business process, although not the formal business process managers), and which were highly successful groups, but no instance where the leader's business process involvement was low and still the group succeeded.

For the purposes of our discussion, as mentioned above, the business process owner is the formal manager of the business process. In other words, owning a business process here means to be the principal coordinator of the several activities that make up the business process, or the one who has the main responsibility for the business process throughput, cost, outcome quality, and other attributes in connection with the business process' execution.

Based on my experience, it is likely that a person in a management position will own at least one business process, but this may not always be the case. For example, some very senior managers do not own any particular business process, even though they may directly influence how a large number of business processes are conducted.

Does this discussion on leadership factors mean that only managers can successfully lead e-collaboration technology-supported business process improvement groups? The answer obviously is *no*. If the answer were *yes*, it would contradict at least one of the findings discussed in previous chapters, which is that e-collaboration technology support decentralizes, from a management perspective, business process improvement initiatives. The evidence shows that a business process owner does not necessarily have to be a manager for a group to succeed. For example, the group whose target business process was teaching a university course had as its leader a part-time graduate assistant, and was nevertheless a successful group.

It is important to note, however, that for a business process owner not to be a manager, the business process in question must be reasonably narrow in scope. And improvements in a business process that is narrow in scope are unlikely alone to lead to radical gains from an organization-wide perspective. This type of business process improvement is what early proponents of business process reengineering used to call incremental improvements (Hammer & Champy, 1993; Kock & Murphy, 2001), and whose individual impact on the organization is likely to be relatively small.

However, the combined impact of incremental business process improvements has the potential to be very bottom-line-significant, provided that they are obtained for a large number of business processes in a more or less synergistic way. The experience of Japanese manufacturers with incremental improvement initiatives in the period from the 1950s to the 1980s shows that this can be achieved with spectacular aggregate results. A number of similar experiences in the US after the 1980s also point in the same direction (Walton, 1989, 1991).

Membership Factors

Table 8.2 shows the variation of membership-related variables among the 12 groups, as well as the degree of success obtained by each group. Groups are sorted by their degree of success, with the most successful being shown at the top, and the least successful at the bottom. The three first columns of Table 8.2 are the same as in Table 8.1. The fourth, fifth and sixth columns show the member attributes that emerged in my interviews with business process improvement group members as the most important, or relevant, member attributes in the context of e-collaboration technology-supported business process improvement. The fourth column from the left shows the member perceptions trend for each group regarding discussion risk; that is, the degree to which careless e-collaboration technology-mediated contributions to the discussion were perceived by members as likely to negatively affect their future careers. The two last columns show the members' general interest in the results of the business process improvement attempt, and the members' direct involvement with the business process targeted for redesign, respectively.

A careful inspection of Table 8.2 suggests that the two membership-related variables that seem to consistently affect group success are perceived risk and interest in business process improvement. With the exception of the group targeting the business process in connection with international graduate student support at Waikato University, the five most successful groups had a low perceived risk and a high interest in business process improvement. The variable business process involvement was never low in any of the groups, moving somewhat erratically from high to medium among the 12 groups, although it seemed to correlate, at least to a certain extent, with the members' interest in business process improvement. Thus, no conclusions can be made regarding the relationship between the members' involvement with the targeted business process and group success.

This means that e-collaboration technology-supported business process improvement groups are likely to be more successful if the perceived participation risk is low and the members' interest in the results of the business process improvement attempt is high.

Table 8.2. Membership factors impacting group success

Process Description	Organi-zation	Success	Members' Perceived Risk	Interest in Process Improvement	Process Involvement
Newsletter editing	MAF	High: Group agreed on process changes, from which all were implemented with positive results	Low: Problems and sources known to all members	High: Process problems affected all members	High: All members played roles in the process
University course	Waikato	High: Group agreed on process changes, from which all were implemented with positive results	Low: Problems and sources known to all members	High: Process problems affected all members	High: All members played roles in the process
Student assignment handling	Waikato	High: Group agreed on process changes, from which all were implemented with positive results	Low: Problems and sources known to all members	High: Process problems affected all members	Medium: Half of the members played roles in the process
International graduate student support	Waikato	High: Group agreed on process changes, from which all were partially or fully implemented with positive results	High: Racial, national origin, and other sensitive issues were discussed	Medium: Two-thirds of the members were affected by process problems	Medium: Half of the members played roles in the process
International student adaptation support	Waikato	High: Group agreed on process changes, from which all were partially of fully implemented with positive results	Low: Problems and sources known to all members	High: Process problems affected all members	Medium: Half of the members played roles in the process
IT users support	MAF	High: Group agreed on process changes, from which three-quarters were implemented with positive results	Low: Problems and sources known to all members	High: Process problems affected all members	High: All members played roles in the process
Software support	MAF	High: Group agreed on process changes, from which two-thirds were implemented with positive results	Medium: Problem sources were unclear	Medium: Two-thirds of the members were affected by process problems	Medium: Half of the members played roles in the process

Table 8.2. Membership factors impacting group success (cont.)

			Members'		
Process Description	**Organi-zation**	**Success**	**Perceived Risk**	**Interest in Process Improvement**	**Process Involvement**
Newsletter editing	MAF	High: Group agreed on process changes, from which all were implemented with positive results	Low: Problems and sources known to all members	High: Process problems affected all members	High: All members played roles in the process
University course	Waikato	High: Group agreed on process changes, from which all were implemented with positive results	Low: Problems and sources known to all members	High: Process problems affected all members	High: All members played roles in the process
Student assignment handling	Waikato	High: Group agreed on process changes, from which all were implemented with positive results	Low: Problems and sources known to all members	High: Process problems affected all members	Medium: Half of the members played roles in the process
International graduate student support	Waikato	High: Group agreed on process changes, from which all were partially or fully implemented with positive results	High: Racial, national origin, and other sensitive issues were discussed	Medium: Two-thirds of the members were affected by process problems	Medium: Half of the members played roles in the process
International student adaptation support	Waikato	High: Group agreed on process changes, from which all were partially of fully implemented with positive results	Low: Problems and sources known to all members	High: Process problems affected all members	Medium: Half of the members played roles in the process
IT users support	MAF	High: Group agreed on process changes, from which three-quarters were implemented with positive results	Low: Problems and sources known to all members	High: Process problems affected all members	High: All members played roles in the process
Software support	MAF	High: Group agreed on process changes, from which two-thirds were implemented with positive results	Medium: Problem sources were unclear	Medium: Two-thirds of the members were affected by process problems	Medium: Half of the members played roles in the process

Figure 8.1. Distribution of perceptions about e-collaboration technology support impact on group participation risk

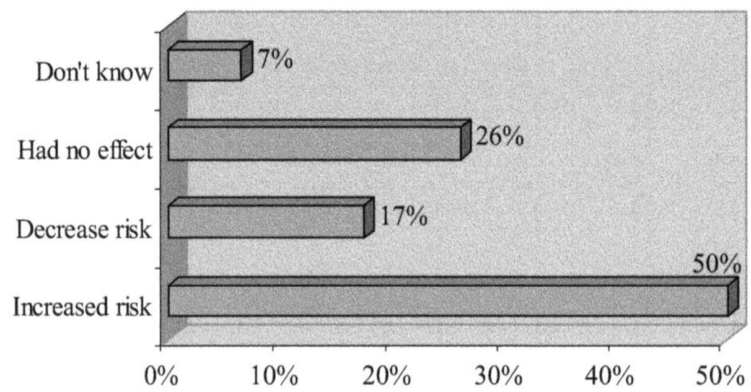

A low perceived risk implies that either the discussion addresses no particularly delicate matters, or that there is enough trust among members so that they can discuss any issue among themselves without fear of potential future backstabbing or negative repercussions of what they said in business process improvement discussions on their future careers.

A high interest in the business process improvement group's outcomes implies that most of those involved in the group have a stake in those outcomes (i.e., most members are directly affected by at least one of the problems identified with the business process targeted by the group).

The perceived risk associated with actively participating in a business process improvement group has been consistently seen as increased by e-collaboration technology support, according to the group members I interviewed. This is evidenced by the trend[2] of perceptions shown in Figure 8.1, which summarizes responses from 46 interviews regarding e-collaboration technology support impact on perceived group participation risk. Fifty percent of the interviewees perceived e-collaboration technology support as having increased participation risk in their groups, while 17% believe the opposite was true; that is, that e-collaboration technology support had, in fact, decreased participation risk.

The main reason given by interviewees to explain why e-collaboration technology support had, in their opinion, decreased participation risk, was the relative isolation experienced when contributing an electronic posting to the business process improvement group discussion. Apparently, these members were basing their explanation on the assumption that in a face-to-face situation, other members could easily and immediately vent their disapproval or anger toward them. In other words, they would be more at the mercy of other members in a face-to-face meeting than in the distributed and asynchronous virtual environment created by the e-collaboration technology.

I find the above mentioned sense of insulation experienced by some individuals in connection with e-collaboration technology mediation in group discussions very interesting, and warranting some theorizing (or philosophizing). I interpret this perception as a trace of a genetically programmed fear of violent reactions and personal confrontations in face-to-face interactions with other human beings. After all, we humans have had to rely on physical strength and the use of violence to survive and procreate during most of our evolutionary cycle. Some people (perhaps timid people) may feel the fear of face-to-face confrontation more intensely than others may. Given that genetically programmed feelings are often subconscious and instinctive, I have some doubts as to whether the insulation perception is rational, or even whether it has anything to do with what can happen in reality in the event of a conflict.

Face-to-face confrontations within organizations rarely lead to physical threats or aggression, and are usually going to be won (though, in many cases, winning is not the most advisable strategy) by the person better prepared to handle an oral argument, without the use of any physical threat or violence. In an e-collaboration technology-mediated confrontation, people may not have to face their interlocutors and, therefore, may avoid immediate unpleasant feelings, but they still will have to face the consequences of their confrontation in the same way. In fact, there is some evidence in e-collaboration research literature suggesting that e-collaboration technology-mediated confrontations can escalate faster and lead to more serious consequences than face-to-face confrontations.

As for those interviewees who perceived an increase in participation risk as associated with e-collaboration technology support, their explanations of their interview answers seemed reasonably logical and rational. The majority of these respondents explained their answers by pointing out that e-collaboration technology-mediated communication leaves a record that can be kept by other group members for future use, and forwarded to many other people. One respondent put it in the following way:

The problem with [e-collaboration technology-mediated communication] is that it is very easy to think of it as a form of conversation, and yet it is also a written record, and people can easily write something down that they later regret. In a verbal (oral) situation, people tend to be a bit more accepting of people saying something inappropriate, and if they say something inappropriate, often the cues of everyone else will protect them.

My interviews, participant observations, and interactions with group leaders suggest that two main situations are seen as particularly risky in business process improvement groups. One of these is the situation where the sources of business process-related problems are not clear. In order to proceed with the group discussion when this situation arises, some members must come forth and either admit that they or their departments (or teams) are causing the problems, or they must accuse other people of being the source of the problems. Either situation is obviously risky, especially since e-collaboration technology mediation usually allows for the recording of whatever these people put forth.

The other risky situation is where classified or sensitive information (e.g., statistics that point to minority-related problems) has to be presented and openly discussed by business process improvement group members. In the case of classified or sensitive information, no group member may want to put forth the information in the first place. If someone does, that may be seen as a violation of organizational norms and regulations (i.e., in connection with privacy and confidentiality issues), or as something that can put the organization in a delicate situation from a legal perspective, given the classified or sensitive nature of the information. These are all obviously risky things to do. In the case of sensitive information, those who provide the information, as well as those who comment on the information, may risk saying something that may be interpreted as inappropriate by others, who can be business process improvement group members or people outside the group who happen to get a copy of the records generated by the e-collaboration system (e.g., electronic postings).

In my opinion, the situations discussed above are likely to arise quite often in business process improvement groups in general. However, I believe that they are more likely to happen in radical than incremental business process improvement attempts. Radical improvement is usually targeted at broad, interdepartmental business processes that typically need urgent attention (Davenport & Stoddard, 1994). A large business process breadth usually leads to several departments being involved in the business process improvement effort. Urgency implies that there are big problems to be solved. The combination of these ingredients yields a business process improvement effort dealing with big problems and involving several different departments. The likelihood that people from different and often competing departments will come forward and state in writing that they admit being at the source of big problems, is, in my opinion, very low (I am talking about normal, not kamikaze-like employees or managers). One may argue that if there is enough trust in the organization, this may happen, but I have not seen this very often in my work with over 100 business process improvement groups (e-collaboration technology-supported and otherwise).

Other Factors

Table 8.3 shows the variation of other variables, also identified by means of interviews, among the 12 groups. Again, as with tables 8.1 and 8.2, the rows in Table 8.3 are sorted by the degree of success obtained by each group, the most successful at the top and the least successful at the bottom. The three first columns of Table 8.3 are the same as in Table 8.1 and Table 8.2. The fourth column from the left shows the degree of complexity of the business process targeted for improvement, which I assessed based on the assumption that it correlates the number of activities of the business process and the perceived complexity of the business process in the eyes of the group members. The fifth column shows the breadth of the business process targeted for improvement, which is proportional to the number of departments or distinct functions involved in the execution of the business process targeted for improvement. The last column shows the group's departmental heterogeneity, which is proportional to the number of different departments represented in the group (i.e., if people from several departments were invited to participate in the business process improvement discussion, then the group's departmental heterogeneity would be high).

No patterns can be inferred from the analysis of the most successful business process improvement groups in Table 8.3 (top half of the table), as there is no predominance of high or low values for any of the three variables among highly successful groups. Yet, when one looks at the bottom half of the table, some patterns clearly can be identified. There is a prevalence of high target business process complexity and breadth among the least successful groups. Four among of the five least successful groups targeted business processes of high

Table 8.3. Other factors impacting group success

Process Description	Organi-zation	Success	Group's Target Complexity	Target Breadth	Departmental Heterogeneity
Newsletter editing	MAF	High: Group agreed on process changes, from which all were implemented with positive results	Low: Simple processes known to all group members	Medium: Interdepartmental	Low: One department (spread across four different sites)
University course	Waikato	High: Group agreed on process changes, from which all were implemented with positive results	Low: Simple processes known to all group members	Medium: Interdepartmental	Low: Two departments
Student assignment handling	Waikato	High: Group agreed on process changes, from which all were implemented with positive results	Medium: Medium-complexity process, one-third of the members knew the process well	Medium: Interdepartmental	Medium: Five departments
International graduate student support	Waikato	High: Group agreed on process changes, from which all were partially or fully implemented with positive results	Medium: Medium-complexity process, one-third of the members knew the process well	Medium: Interdepartmental	High: Eight departments
International student adaptation support	Waikato	High: Group agreed on process changes, from which all were partially of fully implemented with positive results	Low: Medium-complexity process, two-thirds of the members knew the process well	Medium: Interdepartmental	Medium: Four departments
IT users support	MAF	High: Group agreed on process changes, from which three-quarters were implemented with positive results	High: Complex process, only one-quarter of the members knew the process well	Low: Departmental	Medium: Six departments
Software support	MAF	High: Group agreed on process changes, from which two-thirds were implemented with positive results	Medium: Medium-complexity process, one-third of the members knew the process well	Low: Departmental	Low: One department
Pest/disease outbreak communication	MAF	High: Group agreed on process changes, from which one-third were implemented with positive results	High: Complex process, only one-third of the members knew the process well	High: Whole business unit	Low: One department (spread across six different sites)
Undergraduate academic support	Waikato	Medium: Group agreed on process changes, from which none was implemented	Medium: Complex process, all members knew the process well	High: Whole business unit	Medium: Five departments

complexity. Similarly, four out of the five least successful groups targeted business processes whose breadth was high.

Although mainly qualitative and based on a relatively small sample, these patterns become significant when combined with other findings in this chapter

Table 8.3. Other factors impacting group success (cont.)

Quality management consulting	MAF	Low: No agreement on process changes was achieved	High: Complex process, only one-third of the members knew the process well	High: Whole business unit	Medium: Four departments
Staff training and development	MAF	Low: Unfocused discussion, the group was discontinued without agreeing on any process changes	High: Complex process, only one-quarter of the members knew the process well	Medium: Interdepartmental	Medium: Three departments
Student computer support	Waikato	Low: No agreement on process changes was achieved	High: Complex process, only-half of the members knew the process well	High: Whole business unit	Medium: Five departments

regarding leadership and membership factors. They suggest that e-collaboration technology-supported business process improvement groups targeting business processes whose complexity and breadth are high are more likely to fail than are those targeting relatively simple and localized business processes. As with leadership and membership factor patterns, the patterns emerging from Table 8.3 regarding groups' target business process breadth point to incremental, as opposed to radical, business process improvement groups as more likely to be successful when using e-collaboration technology mediation to conduct a large portion of their discussions. Low and medium-breadth business processes are likely to involve either one or a few departments, but not a whole business unit, which makes their improvement less of a reengineering than an incremental improvement exercise.

The interpretation of the high-business-process-complexity-low-group-success trend is not as straightforward as that in connection with business process breadth. The perception that a business process is highly complex does not depend only on the number of activities of a business process. It also depends on how much a person knows about the domain area related to the business process and his or her familiarity with the business process itself. A business process that looks simple to an electrical engineer (e.g., the design of the power infrastructure for a two-story house) may look very complex to a medical doctor. The opposite may be true for the process of operating on an injured human knee. Thus, the evidence suggests, people who share knowledge about a given process are more likely to succeed in improving it in an e-collaboration technology-mediated fashion, even if the process is not that simple. That is, shared process knowledge may make a process look relatively simple to a group, even if the process does not look very simple to someone who does not know anything about it.

This discussion suggests that groups whose members hold shared knowledge about a relatively narrow (i.e., low-breadth) business process are likely to be more successful when improving the business process in a distributed, asynchronous, and e-collaboration technology-mediated fashion. Obviously, these characteristics quite likely are to be found in business process teams who, by definition, work on the same business process and interact face-to-face often. The problem is that for these groups, asynchronous e-collaboration technology mediation has very little use. Unless business process team members work in different shifts, they easily can meet face-to-face, and thus probably will not be interested in meeting electronically.

The groups who can really benefit from distributed and asynchronous e-collaboration technology support are those that involve people from different departments who can rarely meet face-to-face. Those people are unlikely to share much business process-related knowledge, though. As previous chapters show, the involvement of these people in e-collaboration technology-supported business process improvement groups will lead to a buildup in their shared knowledge. In other words, people who engage in e-collaboration technology-supported business process improvement tend to share knowledge, and this knowledge sharing is likely to increase the success of future groups involving the same members. However, until a certain amount of knowledge is shared, business process improvement groups targeting complex (as perceived by most group members) business processes may fail. Therefore, the wisdom apparently lies, as Buddha is believed to have said, in the middle path (as opposed to the extremes). Organizations should stimulate e-collaboration technology-supported business process improvement groups to tackle business processes of medium complexity and breadth, until employees are better equipped (with shared knowledge) to take part in groups targeting more complex business processes that then will no longer look so complex.

Incremental or Radical Improvement?

The previous discussion indicates the following trends regarding the success of business process improvement groups supported by an asynchronous e-collaboration technology.

- First, business process improvement groups where the leader is the owner of the business process targeted by the group tend to be more successful than other groups.
- Second, groups whose discussions present a low perceived risk to their members, and in which most members have a personal stake in the improvement of the target business processes, tend to be more successful than other groups.
- Finally, groups whose target business process is broad and generally seen as complex by group members tend to be less successful than other groups.

So, what does this tell us? For once, it allows us to paint an idealized picture of a successful e-collaboration technology-supported business process improvement group. Such group would have its leader as the owner of the target business process; its discussion would present little personal risk for its members; and most of its members would have a personal stake in its outcome. Such group would also target a relatively narrow and simple business process.

My experience facilitating incremental improvement as well as reengineering groups suggests that this idealized picture fits the incremental business process improvement group picture much better than it fits what is usually known as the reengineering group. The success factor analysis of the 12 groups discussed here, plus some of my previous studies of e-collaboration technology-supported business process improvement (Kock, 2001b; Kock & Corner, 1997; Kock & Davison, 2003; Kock & McQueen, 1995, 1998) strongly suggest, in my view, that e-collaboration technologies are most likely to succeed when used as a support tool for incremental business process improvement groups. These groups should focus on relatively narrow business processes (i.e., business processes that involve a few departments in their execution) and be led by the main person responsible in the organization for those business processes who, I would like to emphasize, does not necessarily have to be a manager.

Can E-Collaboration Technology Support be a Trap?

Although leading to generally positive effects if used in the proper context, e-collaboration technology support can also be a trap under some circumstances. This chapter has discussed success factors, which can also be seen as factors whose absence may lead business process improvement groups to fail. Several of the variables identified as being affected by e-collaboration technology support depend on an initial choice made by the group leader when selecting business process-related problems and group members. For example, a choice of a problem such as "the design of our new products is not properly aligned with our marketing strategies" is likely to lead to a strategic-level choice of a business process. An e-collaboration technology-supported group tackling such a business process is likely to fail if it tries to conduct a large amount of its discussion through an asynchronous e-collaboration technology, because of at least some of the several reasons discussed in this chapter.

Among the myriad decisions a business process improvement group leader may have to make, one specific type of decision can be particularly affected by the availability of e-collaboration technology support. The decision I refer to is the one related to who will participate in the business process improvement group. The reason why e-collaboration technology support can negatively affect this decision is that e-collaboration technology support makes it very easy for anyone to participate in a business process improvement, even a person that has absolutely no interest in the outcome of the group.

The group whose target business process was student computer support at one of Waikato University's colleges, provides a good example of how e-collaboration technology support may become a trap when a self-appointed leader is selecting group members. The self-appointed leader of this group was a graduate assistant who worked in one of Waikato University's academic departments. This individual had previously worked as a computer support consultant, and was also familiar with business process improvement methods. He gained support from the manager of the college's computer support division, who was eager to attain some strategic input from outside of his immediate support staff. He then went on to invite a number of faculty and staff from five academic units to participate in the business process improvement group; some of the invited faculty and staff were at senior-level positions. His main criterion for member selection was to invite people who had computer-related material in their course curricula. Most of those invited to become members agreed to participate immediately, apparently without much thought.

As shown in Table 8.2, this group was the least successful among the entire set of business process improvement groups investigated. The group lasted 32 days. Seven members from the 11 who initially agreed to participate were active contributors to the discussion (i.e., they wrote, as opposed to only reading, electronic postings). According to estimates provided by the group members, 96% of the total time spent by group members in the group discussion was spent in interactions through the e-collaboration system. In the end, the group members achieved no agreement on business process changes, and the group was generally considered a failure from a business process improvement perspective. Some members pointed at the disturbing contributions of one particular group member, a senior faculty member at the college, as one of the reasons why this group failed. The referred group member had, among other things, allegedly addressed other members in a demeaning and sometimes offensive way. Later, in an interview, this member told me the following:

I was a bit naughty, but I had already made my decision that [the group discussion] was not going to be effective, so I felt it was not going to be so much of a loss anyway. So, I basically, quite deliberately, upped the stakes by using phrases and language which were very exclusive, and quite controversial ... It was my way of saying: "You guys need to get a life, we need to move on because this is not going to work." It was the ultimate form of arrogance, if you want. I was playing a game.

During our interview after the group was concluded, this particular business process improvement group member declared having absolutely no interest in the improvement of the business process itself, as he was not involved in the business process in any way; neither as a business process client nor as a member of the team responsible for performing the business process. He had decided not to be a client of the business process, as several of his peers had, because he thought of himself as able to carry out the business process activities himself. That is, he felt that he was able to take care of any computer-related problems he might have himself. This was itself an indication that he was unhappy with the role the computer support division had been playing at his college. As an unhappy customer, he was initially seen as possibly a source of valuable input to the business process improvement group. This obviously turned out to be a misassumption.

The above example shows that the inclusion of some members, who would otherwise not participate in the business process improvement group, may be facilitated by e-collaboration technology support. Although apparently beneficial, this also may become a trap for well-intentioned group leaders. Different from e-collaboration technology-supported groups, face-to-face meetings may be disruptive to their members' personal schedules, may force some of them to travel long distances to participate, and may lead some of them to have to cancel other important appointments. That is, there is a personal cost to be met by those that accept to take part in a face-to-face business process improvement group. Such cost may put off prospective members who do not have a high personal stake in the outcomes of the business process improvement group, and thus induce a natural selection of the most interested individuals for the group. As discussed before in this chapter, a high personal interest in the outcomes of a business process improvement group is likely to contribute to the group's success. E-collaboration technology support often makes it initially very easy for people to participate in business process improvement groups, which can lead to the inclusion of undesirable members.

Summary and Concluding Remarks

There is strong evidence that the current trends toward business process improvement and organizational learning are going to carry on for quite some time, perhaps a few decades or more. We can also observe a trend towards what is known as the virtual organization paradigm. This picture implies that both business process improvement and knowledge sharing will increasingly have to be carried out in a distributed, asynchronous, and e-collaboration technology-mediated manner.

Previous chapters discussed the relationship between business process improvement and knowledge communication, which is one of the main components of organizational learning. The impact of asynchronous e-collaboration technologies on business process improvement and knowledge sharing was also discussed. What was missing in previous chapters of this book was a discussion of success and failure factors in e-collaboration technology-supported business process improvement groups, which has been provided in this chapter.

Three main groups of factors seem to be associated with success and failure in e-collaboration technology-supported business process improvement groups. These are leadership, membership, and other identified factors that are unrelated to leadership and membership. Leadership factors relate to characteristics of the leaders (or moderators) of e-collaboration technology-supported business process improvement groups. Membership factors relate to group membership configurations. As for the other factors, those relate to general characteristics of each group, including characteristics of the target business process.

Among leadership factors, the most closely related to group success was the business process involvement of the group leader. That is, the more deeply involved the group leader was with the execution of the business process targeted by the group, the better the chances seemed that the group would succeed.

Among membership factors, the most closely related to business process improvement group success were perceived risk and business process improvement interest. That is, the most successful groups appear to be those in which there is little risk associated with active participation (as perceived by the majority of group members), and those groups in which most members are directly affected by business process problems (and who thus have a personal stake in the improvement of the business process targeted by the group).

Finally, among the other identified factors, the most closely related with business process improvement group success were target business process complexity and breadth. There seems to be a prevalence of high target complexity and breath among the least successful groups.

The analysis of success factors allows us to paint an idealized picture of a successful e-collaboration technology-supported business process improvement group. Such a group would have its leader as the owner of the target business process; its discussion would present little personal risk for its members; and most of its members would have a personal stake in its outcome. Such a group would also target a relatively narrow and simple business process.

The ideal picture painted here fits the incremental business process improvement group much better than what is usually known as the reengineering group. Organizations should keep this in mind when embarking on e-collaboration technology-supported business process improvement efforts.

Endnotes

1 Or business processes, as some groups targeted more than one business process. Later in this book, more details are provided on each of several individual business process improvement groups, including details about the business process or processes targeted for redesign.

[2] A Chi-squared test indicates that the probability that this trend is due to chance is around 1.5%, which also suggests that the trend is statistically significant.

Chapter IX

Some Realistic Recommendations for Organizations

Information and Knowledge Explosion

The first electronic digital computer, the ENIAC, was developed in 1946 at the University of Pennsylvania in Philadelphia, with funding from the United States Army. By then, computers were seen as giant calculators, capable of performing thousands of complex mathematical operations per second. As World War II had just ended, and the prospect of a global nuclear race was looming large, one of the main applications of computers at that time was ballistics calculation. Among other ballistics-related applications, computers were extensively used for the calculation of warhead missile trajectories with both high speed and unprecedented precision.

This situation slowly changed in the 1960s and 1970s, with the development of smaller and cheaper computers and the first attempts to set up large area networks linking several computers. This period saw the development of ARPANET, the precursor of the Internet, by the US Department of Defense. One of the components of ARPANET, and by no means the most important component at the design time, was an application then called "electronic mail" (or simply e-mail). In spite of the fact that ARPANET's main goal was to allow scientists to spread throughout the US to share computer resources (mostly mainframe computing power), the single most successful of ARPANET's components was, surely enough, e-mail.

This event was a first hint that information and knowledge sharing were becoming more pressing needs than the distributed use of computer processing power. Computers were starting to be seen as communication support and data storage tools, rather than just incredibly fast number crunching devices.

The 1980s saw the widespread use of microcomputers, particular IBM PC compatibles, and the first successful commercial network operating systems. As previously mentioned in this book, one of the most successful companies in the development of local area network operating systems at that time was Novell Corporation, a company based in Utah. Novell's Netware operating system presented two basic attractions to organizational customers.

The first attraction was the ability to allow several users to share what then were relatively expensive computer peripherals, such as laser and color printers, over the network. The second attraction of Novell's Netware was to enable data sharing among users, which meant that users were able to share business process-related information and software applications over the network. The Netware operating system allowed data and peripherals sharing in an almost seamless way. It carved a new information technology market and opened the way for a number of other similar products.

Computer networks became ubiquitous in the 1990s with the emergence of new and cheaper network operating systems. A major player in the stand-alone operating systems market, Microsoft Corporation, launched three operating systems with built-in networking capabilities for a fraction of Novell Netware's price. These operating systems were dubbed Windows for Workgroups[1], Windows 95, and Windows NT. As a result, computer networks became commonplace at organizations of all sizes. These were major steps in connecting computers and, therefore, people through local are networks (LANs). However, LANs were made up of computers located usually in the same building, which limited the scope of applications that could be run effectively on them.

A major breakthrough came with the development and widespread use of the Internet, largely propelled by the development of the World Wide Web in 1991, the first Web browser in 1993, and the emergence of Internet service providers (ISPs). ISPs provided relatively easy access to the Internet, initially to non-governmental companies and individual users. With the advent of the Internet, physical location was no longer an issue, and it became easier for organizations to set up wide area networks (WANs).

This picture is not complete without a brief discussion about the explosion of information and knowledge available for the production and delivery of goods and services (see the introductory chapters of this book for a more detailed discussion). Information is eminently descriptive. It describes the world as it is, as well as changes in its components (e.g., the weather, the economy, a particular organization). Therefore, two main sources of new information are new discoveries about the world that we live in[2], as well as changes in the world. There is no general agreement about how much information is generated every year, but since the pace of new discoveries and global change both have been accelerating at least since the early 1900s, it is reasonable to assume that more and more information is created every day; that is, the rate of information generation has been increasing over time.

New discoveries mean new knowledge, in addition to new information, and as the pace of research increases in different fields, so does the pace with which knowledge is created in these fields (we are, of course, assuming that research usually leads to new knowledge, but not that all research accomplishes that). At least one of my previous studies shows, beyond much doubt, that the number of different experts involved in carrying out a business process is proportional to the number of information exchanges in the business process (Kock et al., 1997; see also Kock, 2003). That is, knowledge fragmentation correlates information flow. Also, as previously discussed in this book, knowledge fragmentation into different fields of expertise is a direct consequence of the explosion in the amount of existing knowledge, as human cognitive limitations lead human beings (who cannot store too much knowledge about too many things in their brains) into knowledge specialization.

From the discussion, it is fair to conclude that there is a pressing need for large amounts of information to be exchanged among people in organizations, and that this need has been increasing over the years. It is also reasonable to conclude, mostly based on the historical discussion of the use of computers, that computer networks are being, and are likely to continue being, the basic infrastructure through which information is exchanged[3]. In this environment, organizations will have to muster techniques to make distributed, e-collaboration technology-mediated work (e.g., routine business process execution) happen in an efficient and effective way. Moreover, as organizations will have to continuously improve themselves and adapt to change, this will also have to be carried out in a distributed, e-collaboration technology-mediated fashion. Below I present and discuss some approaches to how this can be done, based on the arguments presented in previous chapters.

Distributed Improvement and Learning

Previous chapters presented a relatively optimistic view of the effects of e-collaboration technology support on distributed business process improvement groups. The range of effects discussed can be placed into three main categories—group process effects, group outcome effects, and effects on knowledge sharing.

Group process effects refer to the process (or meta-process) of conducting distributed business process improvement groups, and the related effects of e-collaboration technology support. The evidence presented and discussed in previous chapters indicates that e-collaboration technology support causes a reduction of group setup time and group demand for senior leadership. A lower group setup time leads to a lower group duration, or lifetime; that is, business process improvement groups can be completed faster. Finally, the combination of a decrease in group duration and in group demand for senior leadership leads to an increase in the possible number of simultaneous groups that can be conducted at any organization at a given time.

As for group outcomes, the evidence discussed so far in this book is generally inconclusive regarding significant positive or negative effects. The evidence indicates a generally neutral bending toward a positive and certainly non-negative trend in members' perceptions regarding business process improvement group outcome quality. This is due to three main primary effects of e-collaboration technology support. The support provided by the e-collaboration tool appears to increase group departmental heterogeneity, which has been measured as the number of different departments represented in a group. The support provided by the e-collaboration tool also appears to increase individual member contribution quality, perceived as the general quality of individual e-collaboration technology-mediated contributions in a group. E-collaboration technology support is also seen as being associated with an increase in group discussion ambiguity, or the likelihood that members of a business process improvement group would misunderstand individual e-collaboration technology-mediated contributions made by other members.

The combination of the primary effects seems to lead to a neutral effect on group outcome quality, but that doesn't seem to be the case with knowledge sharing. The same primary effects discussed here appear to lead to an increase in knowledge sharing. Two types of evidence support this finding. The first type is perceptions of group members (provided in interviews) about the individual learning they experience in e-collaboration technology-supported groups in comparison with similar face-to-face groups. The second type of evidence in support of the notion that e-collaboration technology mediation leads to increase knowledge sharing comes from the content analysis in connection with the number of knowledge exchanges that take place in improvement and routine business processes. As it has been shown earlier in this book, knowledge is exchanged to a much greater extent in improvement than in routine processes, an effect that seems to be reinforced by e-collaboration technology support.

As a whole, the findings just discussed are generally positive. They are summarized in Figure 9.1, which provides an integrated and summarized view of the many e-collaboration technology support effects discussed in this book. The bottom-line effects are shown on the right-hand side of the figure, which includes the effects on the main dependent[4] variables identified in my past research in connection with the effects of e-collaboration technology support on group-based business process improvement. The dependent variables are the number of possible simultaneous groups and group-based knowledge

Figure 9.1. An integrated view of e-collaboration technology support effects on distributed business process improvement groups

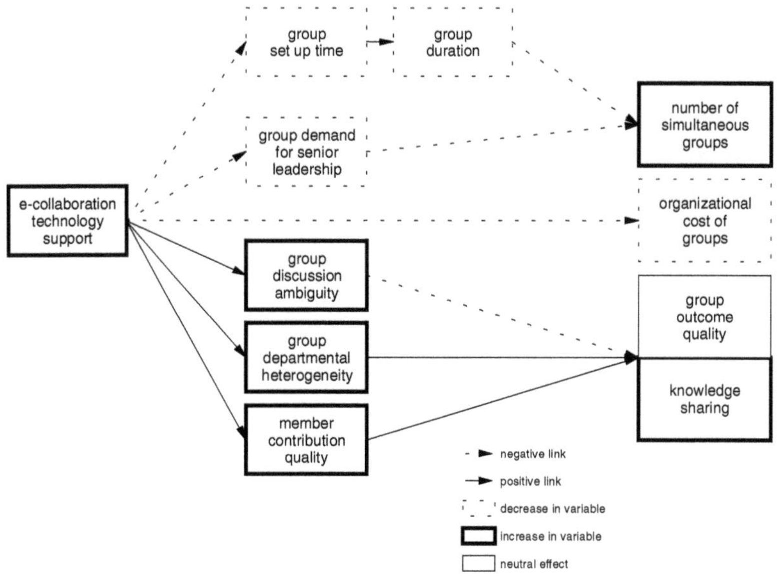

sharing, both of which seem to be increased by e-collaboration technology support. Also, a main dependent variable is the organizational cost of groups, which appears to be decreased by e-collaboration technology support. The fourth and final dependent variable is group outcome quality, which seems to be unaffected by e-collaboration technology support.

Primary and secondary effects on intervening variables mediate the effects on the main dependent variables. Arrows connecting variables in the models represent these effects. All of the other variables in Figure 9.1 are intervening variables, with the exception of e-collaboration technology support, which is an independent variable.

In Figure 9.1, a dotted arrow represents a negative effect; that is, one in which an increase in the variable on the left contributes to a decrease in the variable on the right. A solid arrow represents a positive effect; that is, one in which an increase in the variable on the left contributes to an increase in the variable on the right. Variable changes due to causal effects are represented through the border of the rectangles in which the names of the variables are contained. A thick border means an increase in the variable; a dotted border means a decrease; and a regular border means that the variable is unaffected.

It is very important, however, that the model in Figure 9.1 be interpreted in light of the group patterns discussed earlier in this book regarding success and failure factors in distributed e-collaboration technology-supported business process improvement groups. The findings summarized in the causal model suggest that more simultaneous business process improvement groups will be conducted if distributed e-collaboration technology support is available, and that conducting those groups will be less expensive for an organization than conducting similar face-to-face groups. The model also suggests that group outcome quality is not going to be affected and that knowledge sharing will be increased.

However, whenever a business process improvement group fails, these gains are not actually realized; at least not to their full potential. Who cares if business process improvement becomes cheaper because of e-collaboration technology support, if it is unsuccessful? If this is the case, it may be better not to run a business process improvement group in the first place, unless substantial knowledge sharing gains are achieved, and those gains outweigh the related losses associated with running the group. Also, there is no clear evidence that the amount and quality of learning experienced by group members can offset the losses (i.e., wasted time) caused by an unsuccessful business process improvement effort.

In fact, my own direct observation of business process improvement groups indicates that, with very few exceptions, group failures are seen by most group members as simply failures, irrespective of how much members learn as a result of the group discussion. The main reason for this is probably that business process improvement groups are initially set up to accomplish business process improvement, not knowledge sharing. If the main goal, which is business process improvement, is not achieved, group members more often than not will see their group as a failure. Consequently, those group members may not want to take part in future e-collaboration technology-supported business process improvement groups.

Therefore, management must take some precautions when promoting e-collaboration technology-supported business process improvement in their organizations, despite the generally positive impact of e-collaboration technology support on distributed business process improvement groups. Such precautions are presented and discussed below, organized as a set of prescriptions for organizations.

Some Recommendations for Organizations

There is a large number of organizational recommendations that can be derived from what was discussed in this chapter and in previous chapters of this book. But, since the reader of this book is probably more interested in a synthesis at this stage than in another dose of analytical arguments, I will be as brief as possible without lacking much completeness. I believe my prescriptive comments can be summarized into three main items.

- The leader of a business process improvement group should be deeply involved in the execution of the target business process. Ideally, he or she should be the business process owner. This does not mean that the group leader should be a manager, but it implies that if the leader is not a manager, the scope of the target business process should be relatively limited. This feature more likely is to be found in incremental business process improvement groups than in reengineering groups.

- E-collaboration technology-mediated discussions should not be used alone in broad-scoped, radical business process improvement groups (e.g., reengineering groups). Because of the high level of perceived risk and business process complexity involved in such groups, face-to-face meetings are a must for effective completion of the groups. E-collaboration technology support can be used to conduct part of the group discussions, particularly to summarize and provide a record of what was discussed face-to-face.
- Only people who have a stake in the outcomes of a business process improvement group should be invited to participate in the group. Group leaders should make sure that members whose interest in business process improvement is only marginal are not included in the group. The same is true for members who do not want business process changes to take place and who could, therefore, sabotage the group discussion. This is particularly important in e-collaboration technology-supported groups, as there is evidence suggesting that disruptive or undesirable behavior is more difficult to handle electronically than face-to-face.

In my opinion, most e-collaboration technology-supported business process improvement groups will be successful, if the above recommendations are followed. This will occur without major losses in the benefits of e-collaboration technology support. I assume that a group methodology similar to the one discussed later in this book (i.e., MetaProi) is followed, which involves a number of further normative components (e.g., criteria, prescriptions, guidelines, graphical tools).

These recommendations do not involve major modifications in organizational culture. This is probably a major advantage of these recommendations over the broader culture-oriented prescriptions so often found in the popular management literature. Changes in organizational culture are notoriously difficult to implement successfully. However, it is useful to examine how the creation of an organizational culture that is conducive to e-collaboration technology-supported business process improvement and knowledge sharing could be accomplished.

Popular Beliefs and Not So Popular Realities

Popular business literature is filled with suggestions implying that organizational cultures that promote trust among their members are the ideal ones to stimulate business process improvement and knowledge sharing. Case studies are presented portraying some companies as having unbelievably trusting and friendly environments, apparently devoid from any sign of internal animosity or competition. These companies are often presented as the most successful in their industries, a paragon to be venerated and imitated. Many readers are so marveled at these stories that they become priests of the new gospel, trying to replicate the same environment in their own organizations. I must admit that I was one of such converts a while ago, when I first heard about some of the ideas proposed by William Deming, the father of the total quality management movement, and Tom Peters, a well-known management consultant and best-selling business author.

However, my own experience often has contradicted the idea that organizations can be turned into extremely safe, friendly, and trust-oriented environments, even when top management encourages this culture through incentives and personal example. The problem seems to have its roots in the fact that we live in a competitive environment, which is taught to us in our early years and throughout our lives. Kids compete for grades in school, for prizes in local sport competitions, and even for the right to be left alone by other bullish kids. Could this be any different? After all, we are animals that evolved through Darwin's natural selection, itself a highly competitive process aptly called "survival of the fittest." It is reasonable to expect that this is in some way ingrained in our genetic code as a species, and that this guides our own behavior to a large extent.

Frankly, I think it is a bit hypocritical to try to present organizations as extremely trusting environments. Perhaps some organizations are like that, but most aren't and will never become that way. This reminds me of a situation that I witnessed several times in different circumstances. An executive tells an organization's employees to help build a trusting and friendly environment, because that is very important for the organization to succeed… and those who don't help will be fired!

Other organizational development approaches exist that propose notions that are different from that of creating unbelievably trusting and friendly environments, but that seem to be equally impossible to achieve in practice. The notions proposed by those approaches seem to exist only in the minds of their authors, who, nevertheless, quite often become successful self-proclaimed management gurus.

For example, there is the argument that we should avoid the fragmentation of knowledge and work, which, some argue, perniciously pushes people into becoming part of common-interest groups in organizations and isolating themselves from people outside these groups. Sometimes I think that those who propose these and similar approaches are very smart, because they probably know that they never will be proven wrong. Since it is impossible to avoid knowledge fragmentation and work specialization, unless we find a way to stop the seemingly unstoppable knowledge and information explosion, the organizational cultures prescribed by these so-called management gurus will never be fully implemented. When faced with failures to prevent knowledge and work fragmentation, the gurus always will be able to say, as many of them do, "You haven't done it quite right…not exactly as I said." After all, if what someone is saying never can be tested, then the person never can be proven wrong.

I guess sometimes it pays off to study philosophy and use that study as a basis to come up with management ideas. For example, philosopher Karl Popper pointed out something very similar to the above several years ago (Popper, 1992), when he stated that a theory is only worth considering if it can be tested and possibly proven wrong. This is known as Popper's falsifiability criterion. I think he was absolutely right in that respect.

Having said the above—okay, having gotten it off my chest—it is also important to say that some things can be done to promote positive changes in an organization's culture. More specifically, I do believe that some things can be done to promote a culture oriented toward business process improvement and knowledge sharing. However, we have to be as realistic as possible and first accept some organizational realities.

First, most people will keep on competing aggressively for everything that is even remotely associated with monetary rewards, particularly in countries that promote competition as a way of life, like the US and many others—even Russia and China, whose conversion to capitalism was done with a vengeance. Thus, organizations will have to figure out alternatives to use individual competitive drive to promote synergy among separate organizational functions.

Interestingly, some organizations seem to try to promote competition among their employees, which often ends up backfiring, because people become too stressed, and the quality of what they do eventually goes down. The truth is that most organizations do not have to push people into being more competitive than they already are. They would be better off easing the demand for competitiveness. Releasing pressure on individuals will probably reduce individual stress and increase productivity without any curbing effect on competitive behavior.

A second organizational reality that we should learn to accept is one in connection with knowledge and work fragmentation. The trend towards knowledge and work fragmentation is an irreversible one, and will continue its accelerated march into the future. Therefore, organizations will have to find ways to bring experts together in collaborative efforts, as opposed to trying to turn everyone into a generalist.

Finally, a third organizational reality that we must accept is that computer technologies will be ubiquitous. Most computers will be networked, whether they are within or outside traditional organizational boundaries. This will lend new meaning to the term virtual organization. Practically anyone who has access to the Internet from home can also reach (and be reached by) work colleagues at home. Telecommuting and working from home will become more and more common. As e-collaboration technologies evolve, particularly regarding more natural computer interfaces, geographical organizational barriers will be less and less an issue in the implementation of cross-departmental business processes. Therefore, companies will have to learn how to effectively implement and use e-collaboration technology support tools for distributed work.

Organizational Culture Transformation through Education

Organizational problems, or illnesses, have been traditionally attacked using approaches that resemble those of conventional medicine. That is, organizational "doctors," usually management consultants, hand down prescriptions to managers and employees, who are expected to follow those prescriptions so that the problems can be eliminated.

However, some clinical fields adopt a different approach. In these fields, the first step that clinicians take to fight body illnesses is to educate the patients about their illnesses. By transferring specialized knowledge to patients, doctors help patients to help themselves and thus prevent the recurrence of former habits that were previously contributing to the disease. This is particularly true for illnesses that are caused by complex brain processes, whose chemical dynamics are not yet very well understood. Clinical psychologists and psychiatrists, for example, long have found that some psychological disorders can be effectively treated through educating their patients about the cause of the disorders, with little or no chemical treatment.

In a way analogous to using education to help cure psychological illnesses, I believe education can be used to improve organizational norms, habits, and culture. Underlying this belief is the idea that education provides the basis on which organizations can effectively self-organize themselves, sometimes in completely different ways, in response to market pressures. This view is similar to that put forth by Maturana and Varela (1991) in their theory of autopoiesis, which has been extended to explain behavior in social and organizational systems (Mingers, 2002).

Okay, but education about what? The usual education that an organizational member is likely to undergo is, more often than not, in connection with that person's main job responsibilities, as well as with generic human resources matters. Beyond that, there are several items that follow directly from the discussion presented earlier in this chapter and in previous chapters.

For starters, people should understand the nature of their aggressiveness and competitiveness, and how it can be harnessed to produce positive individual and organizational results. Also, organizations should also promote management and employees' understanding of such concepts as business process and knowledge. The trend towards knowledge fragmentation and its organizational consequences also should be understood. Based on this, efficient approaches should be devised to convince people that they should know something about what others do, particularly business process teammates located in different departments, and the internal and external customers of their business processes. Finally, prospective members of business process improvement groups should understand how e-collaboration technology support could affect their groups.

Summary and Concluding Remarks

Previous chapters discussed several issues regarding the impact that e-collaboration technologies have on business process improvement and knowledge sharing. These can be briefly summarized as follows:

- An increase of the number of simultaneous business process improvement groups that can be conducted at an organization.
- A decrease of the organizational cost of business process improvement groups.
- A neutral effect on business process improvement group outcome quality.
- An increase in knowledge sharing, which is reflected in the perceptions of members of business process improvement groups.

However positive the above findings may sound, they can only become reality if some precautions are taken by organizations. These can be summarized in the three following guidelines:

- Business process improvement group leaders should be deeply involved in the execution of the business process targeted by their groups. Ideally, they should be the owners of those business processes. Leaders, however, do not need to be part of the management staff at organizations.
- E-collaboration technology-mediated discussions should not be used alone in broad-scope, radical business process improvement groups (e.g., reengineering groups). E-collaboration technology support can be used to conduct part of the group discussions in these cases.
- Only people who have a stake in the outcomes of a business process improvement group should be invited to participate. This is an issue of particular concern, because e-collaboration makes it apparently very easy for anyone to participate in a business process improvement group, which may lead to the inclusion of undesirable (e.g., disruptive and/or disinterested) members.

Finally, e-collaboration technology-supported business process improvement and knowledge sharing will be accomplished best if they are rooted within an organizational culture that is conducive to self-analysis, continuous improvement, and learning. Such a type of culture can be built by educating organizational members about business process improvement's mechanics and their likely results. Efficient approaches should be devised to convince people that they should know something about what others do, particularly business process teammates and customers. Finally, prospective members of business process improvement groups should understand how e-collaboration technology support could affect their groups.

Endnotes

1 Many saw Windows for Workgroups as an operating "environment" rather than a full-fledged operating system. This was because it needed, as its predecessor (the stand-alone Windows), the Microsoft DOS operating system to run. Windows for Workgroups carried out most of its functions by making system calls to Microsoft DOS.

2 These include discoveries about us, our body, and our mind, and about how we behave individually and in groups.

3 Note that I am not talking here about *knowledge* exchanges, because I do not think that much knowledge is, in fact, exchanged in current organizations, or that current information technologies contribute much to knowledge communication. Obviously, I think they should, and this is one of the main points I have tried to make in previous chapters of this book.

4 Dependent variables are the most important variables in a cause-effect model such as the one discussed here. The effects on these variables usually summarize whatever causal relationships exist in a model.

Chapter X

Using MetaProi to Improve Business Processes

MetaProi at a Glance

In my earlier discussion in this book about business process improvement and organizational learning, I have shown that business process improvement has the potential to foster interfunctional knowledge communication and, consequently, organizational learning. In previous chapters of this book, I have analyzed business process improvement efforts that led to levels of knowledge communication not normally seen in routine organizational processes. Those business process improvement efforts have all been carried out through business process improvement groups.

Given the potential advantages for organizations from conducting business process improvement groups, the issue of how to conduct such groups becomes very important. There are a number of how-to texts on business process improvement. Such texts describe a variety of normative approaches. Classic texts on how to improve business process quality written by Crosby (1980, 1984), Deming (1986), Ishikawa (1986), and Juran (1989) fed the quality improvement fever of the 1980s. Popular texts focusing on the improvement of business process productivity written by Davenport (1993), Hammer and Champy (1993), Hammer and Stanton (1995), and Harrington (1991), fed the reengineering fever of the 1990s.

In this chapter, I propose my own methodology for business process improvement, namely MetaProi, which, by necessity, is based on the texts just described, as well as other related publications. A previous book presents a preliminary discussion of this approach (Kock, 1995a). MetaProi's foci are on both quality and productivity. The methodology has been designed so it can be conducted through e-collaboration technology-supported and face-to-face meetings. One example of an electronic discussion based on MetaProi is provided at the end of this chapter.

MetaProi is a group methodology for business process improvement. One of its components is a group process (or meta-process). As a methodology, MetaProi can be fully defined as a set of activities, guidelines, criteria, and graphical tools to be used by business process improvement groups. Based on my past experience facilitating business process improvement groups using MetaProi, I would suggest group size to be between three and 25 participants, who would play the roles of group leader, facilitator, and ordinary member. The group's main goal should be to identify an organizational process where improvement opportunities exist, and propose changes in order to translate those opportunities into practical improvement.

MetaProi is short for *Meta*-Process for Business *Pro*cess Improvement. It is referred to as a meta-process to indicate that it is a high-level process that describes how business process improvement ideally should be carried out in organizations. MetaProi is made up of three main stages—business process definition, analysis, and redesign. Each stage comprises interrelated activities. In order to define the criteria, guidelines, and tools to be used in MetaProi, it is important to identify the activities in each of the stages, as well as the group roles involved. Group roles in MetaProi are analogous to business process functions in organizations. The activities involved in each of the stages are summarized below.

- *Business process definition stage*
  - Identify problems
  - Identify business processes
  - Select a business process for redesign
- *Business process analysis stage*
  - Model the business process
  - Raise performance information
  - Highlight opportunities for improvement

- *Business process redesign stage*
  - Search for suitable changes
  - Incorporate changes to the business process model
  - Evaluate redesign feasibility

An illustration of MetaProi as a set of interrelated activities is provided in Figure 10.1. Arrows indicating the flow of data briefly describe the outputs and inputs of the activities.

The illustration in Figure 10.1 is a simplification of the real process. The goal of this illustration is to provide a clear yet limited view of MetaProi as a whole. Loops and interactions with members outside the group are not represented, though these are likely to occur in real business process improvement groups. For example, a group may decide, while performing the activity "evaluate redesign feasibility" that it must go back to the activity "search for suitable changes," due to the impossibility of implementing some of the proposed changes. Also, the facilitator of a group targeting a specific business process in

Figure 10.1. MetaProi as a set of interrelated activities

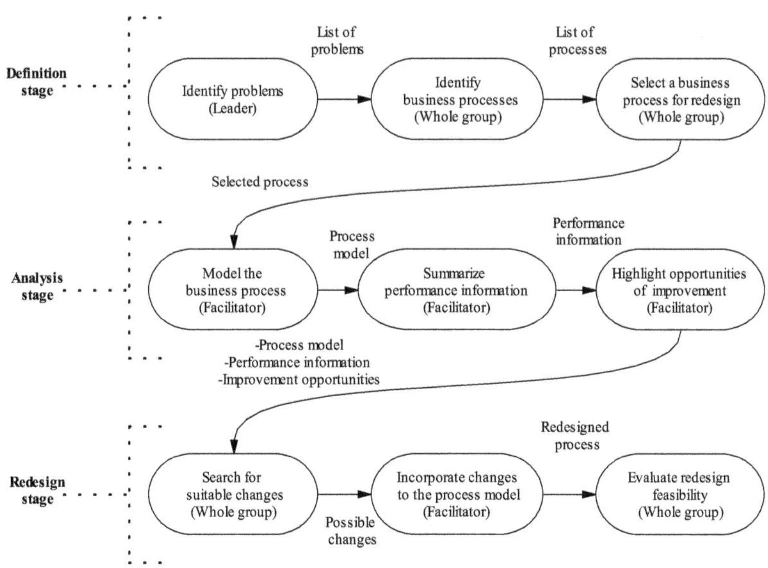

the information technology department of an organization may well need, in the stage "raise performance information," information from the finance department.

Two permanent groups should be set up by an organization implementing MetaProi in order to guarantee the success of business process improvement groups—the Business Process Improvement Committee and the Business Process Improvement Support Team. The Business Process Improvement Committee is usually made up of senior managers, including the chief information and technology officers, and senior management and information technology consultants. The Business Process Improvement Support Team is usually made up of internal information technology support staff, internal business process improvement staff (if a business process improvement department exists, as is the case in some companies such as Lockheed Martin), and management and information technology consultants.

The Business Process Improvement Committee analyzes business process redesign proposals and, when necessary, coordinates and supports their implementation and standardization throughout the organization. Business Process Improvement Committee members should have enough authority to coordinate the implementation of strategic changes, such as those requiring large investments and organization-wide restructuring.

The Business Process Improvement Support Team's main purpose is to provide business process improvement groups with necessary methodological and technological support. The Business Process Improvement Support Team is also responsible for documenting, organizing, and providing public access to the information about business process improvement initiatives in the organization (e.g., documents and electronic postings generated by previous business process improvement groups).

Group Roles in MetaProi

MetaProi comprises only three group roles—leader, facilitator, and member. A business process improvement group is initiated by a self-appointed leader who should initially identify a set of related problems to be tackled by the group. The group leader then invites other members to become part of the group, and appoints one of these members as the group facilitator. The group leader should advise the Business Process Improvement Support Team that the group has been created, so it can support and document the group's evolution.

The leader coordinates the activities of the group and interacts with the Business Process Improvement Support Team. The responsibilities of a group leader include the following:

- Scheduling meetings and making sure the necessary resources are available. Such resources may include a room and overhead projector, or an e-collaboration system.
- Contacting group members and making sure they are able to attend the group meetings, either face-to-face or electronically.
- Gathering and organizing the documentation generated by the group and, after the business process improvement group has completed its work, supplying the Business Process Improvement Support Team with this documentation.

In a business process improvement group, the facilitator is responsible for creating and maintaining a model of the business process targeted for redesign. This model is generated according to one of the business process views discussed earlier in this book.

The facilitator is also responsible for summarizing performance information about the business process, and for highlighting opportunities for improvement. These responsibilities demand a thorough understanding of MetaProi's criteria, guidelines, and tools. However, the facilitator does not decide alone on the adoption of specific changes. This is a prerogative of the business process improvement group as a whole and must be obtained by consensus.

The other members of the group (i.e., the ordinary members) will provide their inputs throughout the group discussion in a relatively low-cost participation mode. As in most types of moderated group discussions, most of the burden is on the leader and facilitator. One person can play more than one role in the group (i.e., one person can be the group leader, the facilitator, and provide inputs as a group member).

General Guidelines for MetaProi

Some guidelines that relate to the whole business process improvement meta-process and which are not associated with a particular activity are the following:

- The business process improvement group should come up with a redesign proposal in a limited amount of time. Based on my past experience, I would suggest that this should be no more than eight weeks. Previous research shows that an acceptable average time is three weeks (Kock & McQueen, 1995).
- The several stages a business process improvement group goes through should be documented. The leader is primarily responsible for this documentation, which is essential to build up historical information about organizational process improvement initiatives. This information can be used for many purposes, such as a basis for future business process improvement groups, and as evidence of the organization's commitment to improving business process quality in accreditation audits (Kock & McQueen, 1997). For example, the organization may use business process improvement group documentation in ISO 9000 accreditation audits to show that it follows exemplary procedures for dealing with non-conformities in business processes (a must in ISO 9000 accreditation, as well as in other types of accreditation).
- Each of the group meetings should be concluded with a link (or hook) to the next meeting. For example, a meeting where the activities "identify problems" and "identify business processes" are accomplished, should end with a preliminary selection of a business process to be redesigned. This preliminary selection works as a link (or hook) to the next meeting, where the first activity will be "select a business process for redesign." These links between meetings are aimed at improving group focus.
- The facilitator should not try to enforce the group process described in this guide (i.e., MetaProi). The facilitator rather should induce it in as transparent a way as possible. This will occur almost naturally, as the facilitator will be responsible for several of the key activities of the business process improvement group.

Activities in MetaProi

The following subsections provide a discussion of each of the activities in MetaProi, including criteria, guidelines, and tools used. Subsection titles are formed by the main stage, which is followed by a colon and the name of the activity.

Definition Stage: Identify Problems

In the definition stage, the first group activity is to identify problems. As discussed before, the person who first brings the problems up for discussion is a self-appointed group leader. Virtually anyone can be a group leader, which helps spread the responsibility for business process innovation throughout the organization, as well as reduce innovation's reliance on managers. This broadens business process improvement's scope of application, as the number of managers in one organization is usually smaller than that of line employees.

In some forms of business process improvement, where the improvement is gradual and accomplished by permanent groups (e.g., quality circles), the search for improvement does not necessarily rely on previous identification of problems. In these cases the improvement is routinely sought, based on the assumption that every business process can always be improved in one way or another. However, research shows that the identification of problems, as sources of discontent within the organization, is a success factor in business process improvement (Hall et al., 1993).

The identification of problems fosters interest in business process improvement among organization members and, at the same time, gives them an idea of what is to be achieved with the improvement. The identification of problems, though, is only the beginning of MetaProi. The main outcome of MetaProi is business process improvement, not problem solving. The identification of problems is an intermediate step that leads to the selection of a business process for improvement (Harrington, 1991; Kock & Murphy, 2001).

A list of interrelated problems first should be generated and then submitted to the business process improvement group so mistakes and omissions can be corrected. The group leader should prepare the preliminary version of the list. This is the first step in the formation of the group. Concurrently with the generation of this list, the leader should invite prospective group members.

Listing problems and inviting group members are two interrelated tasks. Little involvement can be expected from group members who have no interest in the problems initially listed. The problems in the list should be at least intuitively related. A list of problems that is excessively broad, involving several different areas, for example, leads to the identification of several business processes for redesign. This is likely to diffuse the focus of the business process improvement group, and thus should be avoided.

Problems should be approached in a very clear and open way. There should be no fear of disclosing discontent with the actual situation. Poor identification of problems (i.e., certain problems are not discussed because they may upset some individuals) leads to poor business process redesign (Deming, 1986; Kock & Tomelin, 1996; Kock & Murphy, 2001).

Definition Stage: Identify Business Processes

Once a list of interrelated problems is identified, the next step is to identify the business processes causing those problems. At this point, it may be found that some business processes are clearly defined, while others are not (Wastell et al., 1994).

The business process improvement group should not try to build business process models in this activity. Instead, it should try to describe in a few words or sentences the interrelated activities that are perceived by the group as the causes of the list of problems. For example, if a company specializes in performing financial audits, the problems listed may be summarily described as "late invoices," "customer complaints about invoice complexity," "inaccurate invoices," and "late payment." As these are all related to invoicing in connection with the auditing service, the business processes simply can be described in this activity as "invoicing" and "auditing." Later, in the second stage of MetaProi—the business process analysis stage—the selected business process or processes will be analyzed in more detail.

The relationship between problems and business processes is a many-to-many one. Several business processes may cause one problem, and, conversely, several problems may be caused by one business process. Thus, even though the initial list of problems may have only one problem, it may help in the identification of several business processes for improvement.

Definition Stage: Select a Business Process for Redesign

This activity is a conclusion of the work started in the previous activity, the activity "identify business processes." Here, one of the business processes identified in that activity will be chosen for redesign.

When several business processes are identified, group members may be tempted to select more than one business process for improvement. This is frequently the case when there are no clear boundaries between business processes within the organization. However, as the number of selected business processes increases, so does the complexity in the next stage—the business process analysis stage. An additional drawback of a group selecting many business processes for redesign is the high number of changes likely to be proposed by the group. A high number of business processes selected for redesign may hinder the business process improvement group from focusing on one specific business process that needs urgent attention. It may also reduce the level of care given to the analysis and redesign of each individual business process.

Criteria

- The business process improvement group should strive to select as few business processes as possible. Ideally, only one business process should be selected.
- The business process that is associated with the most critical problems should be given priority in the selection.
- After applying the preceding criteria, the business process that is associated with the highest number of problems should be given priority in the selection.

Analysis Stage: Model the Business Process

In this activity, the business process considered for improvement by the business process improvement group is modeled according to one or more of the views discussed earlier in this book. Each view implies a type of business process representation, of which I will discuss here the flowchart (workflow view) and the data flow diagram (data flow view). The goal of this activity is to understand the relationships between business process activities, as well as to achieve a clear view of the business process as a whole.

Graphical Tool: Flowchart

Flowcharts have been extensively used in the past to represent business processes, especially from an information systems analysis and design perspective. Davis (1983) describes a flowchart by defining 16 symbols used as its basic units. Harrington states that: "Flowcharting ... is an invaluable tool for understanding the inner workings of, and relationships between, processes" (Harrington, 1991, p. 86) and discusses five types of flowcharts:

- Block diagram. This is the simplest type of flowchart and uses only two symbols.
- American National Standards Institute (ANSI) standard flowchart. This type of flowchart is more elaborate than the block diagram. Harrington (1991) suggests a list with 12 standard symbols to be used with this type of flowchart (Harrington et al., 1998).
- Functional flowchart. This flowchart uses the same set of symbols as the ANSI standard flowchart. It also includes a horizontal grid describing the functions in the activities as roles.
- Functional timeline flowchart. This flowchart adds some extra information to the functional flowchart, including information about the processing and cycle time of each of the activities.
- Geographic flowchart. This flowchart describes where activities physically take place and how functions and products move within the organization during their execution.

I will describe and use the ANSI standard flowchart. Yet, for the sake of simplicity, I will use only four symbols. These symbols are slightly modified to condense more information than in the ANSI standard. This simplification was adopted in a previous set of studies conducted by me (Kock, 1995a) where several manufacturing and service processes were modeled. The four symbols used are illustrated in Figure 10.2.

The symbols listed in Figure 10.2 are meant to be a minimal list to be used for business process representation. Such simplification has also found wide-

Figure 10.2. Flowchart symbols

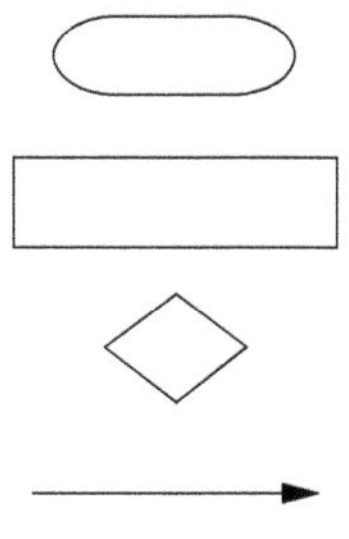

spread acceptance in some organizations, such as Ford and Volkswagen, in business process descriptions for their corporate quality manuals (Kock, 1995a). A discussion of the symbols in Figure 10.2 is provided as follows:

- Start/End: Used to show the beginning and end of a business process. Normally, the words "start" (at the beginning of the business process) or "end" (at the end of the business process) are included within this symbol.
- Activity: Used to represent an activity within the business process. A brief description of the activity is provided within this symbol, together with a description of the function responsible for the activity and the main artifact used (e.g., activity: drill a batch of network cards; function: drill operator; artifact: numeric-controlled drill).
- Decision point: Used to show that at a certain point in the business process a decision must be made. The groups of activities executed after a decision point will vary according to the decision made at the decision point. Typically, decision point symbols are marked with a set of options that describe where the activity flow should proceed after the decision (e.g., yes-no; true-false; option 1, 2, or 3).
- Direction of flow: Used to indicate the direction of the activity execution flow within the business process. Arrows linking pairs of activities indicate the direction of flow. ANSI suggests that the arrow is not necessary when the direction flow is obvious (e.g., from top to bottom).

Figure 10.3. Example of flowchart

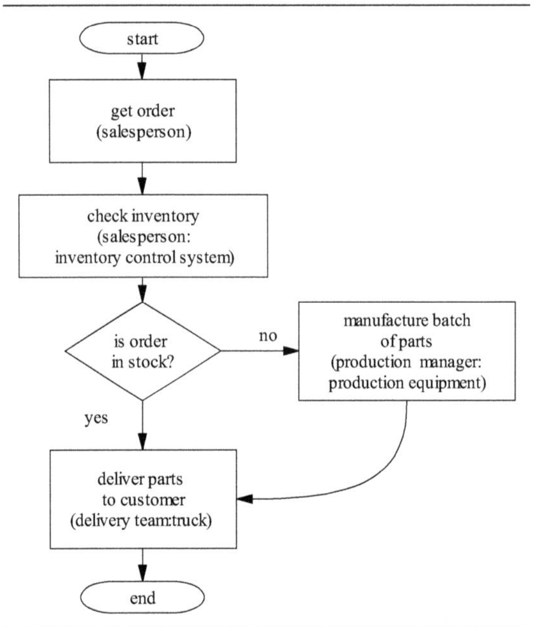

An example of a simple business process modeled with the use of the symbols described above is provided in Figure 10.3. This example was adapted from an illustration in a previous publication (Kock, 1995a) and is aimed at describing a generic order fulfillment process for an automobile part manufacturer.

Each business process flowchart should be accompanied by a written description of the most important activities, highlighting information that may be useful in the business process redesign stage. This description should also include information that is not explicitly represented in the flowchart. For example, it is not clear from the flowchart in Figure 10.3 how the production manager is informed that a batch of parts needs to be manufactured.

As discussed earlier in this book, a flowchart diagram usually leaves out one of the most important components of a business process (i.e., its data flow). Even when the flow of data is included in a flowchart, usually going from one activity or another, that is a misrepresentation of the real data flow in the business process. The main reason is because data do not flow between activities; they flow between business process functions. A type of diagram that is particularly suitable for showing how data flow is the data flow diagram, discussed next.

Graphical Tool: Data Flow Diagram

Data flow diagrams (DFDs) have found widespread use in organizations, particularly due to the widespread adoption of structured analysis and design techniques for the development and deployment of information systems in organizations in the 1970s and 1980s (Davis, 1983; DeMarco, 1979; Kock, 2003; Pressman, 1987). However, DFDs have not been traditionally used to support business process analysis and redesign. Typically, they have been used to understand business processes and automate the processes "as they were." That is, DFDs have been traditionally used for business process analysis, but not redesign (Kock, 2001a, 2003).

Different types of DFDs have been proposed in the late 1970s and widely used since. Notable contributions have been made by DeMarco (1979), Gane and Sarson (1979), and Yourdon and Constantine (1978). The DFDs proposed by these authors differed mostly in the shape, basic function, and number of symbols used in the diagrams. Later, Ward and Mellor (1985) proposed an extended set of symbols to be used in DFDs representing real-time systems (e.g., continuous process control systems).

For convenience, I will describe and use a type of DFD that I have developed by amending a previously suggested notation, and that I have used in most of my consulting and teaching practice. This DFD uses a set of symbols that is not much different from that proposed by Gane and Sarson (1979). The general meaning of the symbols used is virtually the same. The symbol set used is shown in Figure 10.4.

As with the symbols used in flowchart diagrams, I believe the symbol set in Figure 10.4 is enough to provide a nearly complete representation of the flow of data in a business process, with relatively little effort from the business process modeler. The meaning of the symbols shown in Figure 10.4 follows.

Figure 10.4. DFD symbols

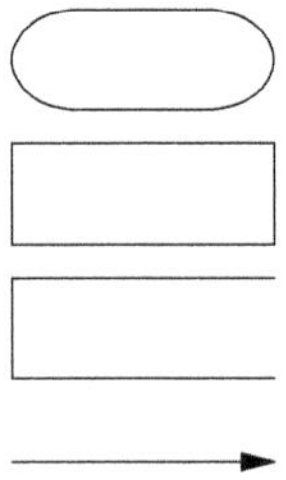

- Activity: Used to represent an activity or sub-process within the business process being modeled. As with the activity symbol in flowcharts, this symbol typically comprises a brief one-to-two-word description of the activity, along with brief descriptions of the organizational function responsible for the activity and the main artifact used.
- External entity: Used to represent a source or recipient of data flowing in a business process. A brief description of the organizational function receiving or supplying data is provided within this symbol.
- Data store: Used to represent a data repository, which may take forms as varied as a computer file or database, an inbox, an e-mailbox, a fax folder, or a billboard. A brief description of that data stored is provided within this symbol.
- Data flow: Used to describe the data components flowing within a business process. As discussed before, data flow between organizational functions, as well as from these to data stores and back. A data flow is described with an arrow pointing to the direction of the flow, along with a brief textual description of its main component.

A simple example of DFD is provided in Figure 10.5. The business process represented is the same as in Figure 10.3 (i.e., a generic order fulfillment process for an automobile part manufacturer). The difference is that Figure

Figure 10.5. Example of DFD

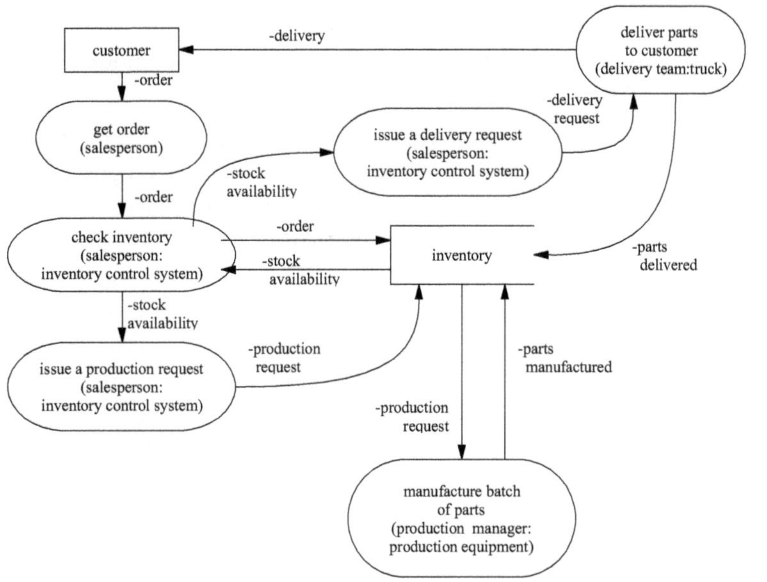

10.3 is a flowchart representation of the business process, whereas Figure 10.5 is a DFD representation.

A comparison of Figures 10.3 and 10.5 highlights some of the usual differences between DFDs and flowcharts. Although the DFD in Figure 10.5 contains descriptions of the same activities (or sub-processes), it also incorporates descriptions about how data flow within the business process modeled. DFDs typically provide a richer representation of business processes that involve data processing (which most business processes do today), particularly in business processes in which a lot of data have to be stored and transferred.

Given the above-mentioned characteristic of DFDs, it should come as no surprise that they are my preferred graphical representation for business process modeling in the context of business process improvement. However, this opinion does not find much agreement among business process modelers, particular those engaged in business process reengineering. A preferred approach among this group is the use of activity-based diagrams, with special attention devoted to the use of IDEF0 representations[1].

Guidelines

- The description within the activity symbols should be as brief as possible and begin with a verb in the infinitive form (e.g., drill a computer card; load a batch of parts onto a truck).
- Flowcharts and DFDs should have a limited number of activity symbols in order to avoid excessive complexity. Studies on human cognition limitations provide the basis for establishing an optimum number of symbols in business process modeling diagrams (Miller, 1956). These studies suggest that this number should be between five and nine grouping symbols (i.e., seven plus or minus two), which, in business process representations, are activity symbols around which other symbols gravitate. When a business process cannot be represented with less than 15 activity symbols (i.e., approximately twice the optimum average) due to its complexity, some of its activities should be exploded (i.e., broken down) into lower-level DFDs (Pressman, 1987).
- Trivial artifacts should not be described in activities (e.g., pen and paper, telephone). A rule of thumb is that one should describe only artifacts that are specific to an activity type, and without which the activity cannot be carried out (e.g., lathe, computerized drill, cheese processor, inventory control system). An artifact is specific to an activity type whenever it has been designed to support primarily that type of activity.
- When modeling a business process, the facilitator should not be afraid to add handwritten notes and marks to the diagram, if they are needed to clarify certain points. The emphasis should be on using the graphical tool in an effective way (i.e., to convey information that will allow the group to successfully redesign the business process) rather than in an efficient way (i.e., keeping the chart as neat and tidy as possible by strictly sticking with the flowcharting symbolism).

Analysis Stage: Summarize Performance Information

In this activity, information about the performance of the business process is summarized for the business process improvement group. This information should revolve around two main business process attributes—quality and productivity. A direct measure of business process quality is customer satisfaction, so the best way to evaluate it is to obtain information on how the customers of the business process perceive its outputs.

The customers of a business process are those inside and outside the organization who receive products generated by the process. These products can be services, goods, information, or computer software. Lathe operators, for example, are customers of a lathe maintenance process. The maintenance service they are provided also affects the quality of the products generated by the lathe operators themselves.

Productivity traditionally has been measured by the ratio outputs/inputs (Misterek et al., 1992). This means that a car assembly process that employs 10 workers and produces two cars per hour may be said to have a productivity of 2/10 = 0.2 cars per worker per hour. If the same business process is redesigned so that it can produce the same two cars per hour, but now with five workers, then its productivity will be 2/5 = 0.4 cars per worker per hour. That is 100% higher than before.

However, a better way to measure business process productivity is by considering the ratio (production capacity)/(production costs). This offers two advantages against the (input/output) approach discussed in the previous paragraph.

- It considers the costs of the inputs to the business process, and not their quantity.
- It takes into consideration the capacity of a business process, and not the realization of that capacity.

The quantity of each input may remain the same, even when its cost is reduced due to business process improvement. For example, the process related to the production of hamburgers can benefit from a smarter purchase of bread, whether the number of bread units is reduced or not. This is why the analysis of cost is critical to productivity measurement, as opposed to the approach of counting the number of inputs. Yet, this approach implies a higher measurement complexity, as costs can vary considerably over time.

The measurement of the production capacity of a business process implies forecasting. To say that a car assembly process has a production capacity of 300 cars a day means that the assembly line can manufacture on average that figure, but not that it is the real average output. Since production in real contexts depends on consumption expectancy, which, in turn, is based on sales orders or forecasts, the simple measure of outputs can lead to wrong assumptions about productivity. This risk is suppressed when productivity assessment is based on production capacity (Goldratt & Fox, 1986; Kock, 1995a; Kock & Murphy, 2001). Again, complexity here is increased by the need to estimate business process output capacity based on historical figures and resource capacity of specific units (e.g., the production capacity of a machine in a manufacturing plant). However, in many cases, this may be easier than relying on real numbers whose measurement is severely hampered by the added cost of extensions in the accounting system of the organization (Mark, 1984).

So, I generally believe that the analysis of productivity should be based on estimates of production capacity and costs, rather than on outputs and inputs. While likely to add complexity to measurement, this is useful in that it draws a line between productivity and quality assessment. The output/input approach disregards the fact that quality improvement is bound to generate more consumption, and consequently promote an increase in output (Deming, 1986). By connecting productivity with the actual outputs produced by a business process, one could mistake quality for productivity improvement. This is particularly true when a surge in demand due to higher quality is simply supported by excess capacity, not augmented productivity.

Guidelines

- In the first activity of MetaProi, the one aimed at identifying problems, the group should have gathered information on user complaints. In this activity, the facilitator should try to find quantitative data associated with those complaints. For example, the facilitator should try to identify by means of quantitative measures, the problems customers see as most critical and those that occur most often.
- In this activity, the facilitator should not be concerned with generating performance information. The facilitator instead should focus on summarizing existing information about the business process performance. This information may come from areas of the organization that are not represented in the business process improvement group. Generating performance information may take too long and, therefore, make the business process improvement group lose momentum. A lack of business process performance information, identified by a group in its analysis stage, may become a problem to be tackled by a different business process improvement group.

Analysis Stage: Highlight Improvement Opportunities

In this activity, the facilitator will highlight opportunities for improvement based on the business process performance information summarized in the previous activity. This is helpful to lead the business process improvement group towards the discussion of concrete changes to improve the business process.

Guideline

The facilitator should highlight business process improvement opportunities by proposing changes in the business process to be discussed by the group. These changes should be based on the information gathered during the two previous activities; namely, "model the business process" and "raise performance evaluation." The group also should follow the guidelines discussed in the next activity; that is, "search for suitable changes."

Redesign Stage: Search for Suitable Changes

In this activity, group members will propose suitable changes in the business process so improvements of quality and productivity can be achieved. The literature on business process improvement provides several guidelines for improving business processes. These guidelines can help business process improvement group members to formulate their redesign proposals.

Guidelines

Harrington (1991) and Harrington et al. (1998) provide several guidelines for business process improvement based on general principles, such as business process and activity simplification, bureaucracy elimination, standardization, and technology utilization. Hall et al. (1993) and Venkatraman (1994) propose guidelines for redesigning business processes according to improvement dimensions and scope levels. Guha et al. (1993) and Wastell et al. (1994) present some business process improvement guidelines as part of specific business process redesign programs. Dingle (1994) and Caron et al. (1994) draw guidelines from the analysis of business process reengineering cases.

In my own work, I often opt for splitting business process improvement guidelines into three main domains of business process improvement—information flow, structure of activities, and management system. I did so as part of an earlier business process improvement methodology named PROI, which stands for Business *Pro*cess *I*mprovement (Kock, 1995a) and that served as a basis for MetaProi. Some of those guidelines that may be useful within the context of MetaProi are listed below.

- Foster asynchronous communication: Whenever people exchange information, they can do it synchronously (i.e., interacting at the same time, or asynchronously; interacting at different times). One example of synchronous communication is a telephone conversation. If the conversation takes place via e-mail, it then becomes an example of asynchronous communication. It has been observed, especially in formal business interaction, that asynchronous communication is almost always more efficient. For example, synchronous communication often leads to wasted time when the communication interaction is being set up (i.e., waiting for the other person to be found), and communication tends to be less objective. Asynchronous communication can be implemented with simple artifacts such as in- and out-boxes, fax trays, and billboards. These artifacts work as dynamic information repositories.
- Eliminate duplication of information: Static repositories, as opposed to dynamic repositories, hold information on a more permanent basis. A student file maintained by an elementary school, for example, is a static repository of information. Duplication of information in different static repositories often creates inconsistency problems, which may have a negative impact on productivity and quality. In one of my previous books (Kock, 1995a), I describe a situation where a large automaker's purchasing division tried to keep two supplier databases updated, one manually and the other through a computer system. Two databases were being kept because the computer database had presented some problems and, therefore, was deemed unreliable; so, the solution devised by those involved was to try to maintain two databases, one as the main database and the other as a backup database. This, in turn, was causing a large number of inconsistencies between the two databases. Each database stored data about more than 400 parts suppliers.
- Reduce information flow: Excessive information flow is often caused by an over-commitment to efficiency to the detriment of effectiveness. Informa-

tion is perceived as an important component of business processes, which drives people to an unhealthy information hunger. This causes information overload (Buchanan & Kock, 2000; Kock, 2000; Toffler, 1970) and the creation of unnecessary information processing functions within the organization. Information overload leads to stress and often the creation of information filtering roles. These roles are normally those of aides or middle managers, who are responsible for filtering in the important bit from the information coming from the bottom of, and from outside, the organization. Conversely, excessive information flowing top-down forces middle managers to become messengers, to the damage of more important roles. Information flow can be reduced by selecting the information that is important in business processes and eliminating the rest, and by effectively using group support and database management systems.

- Reduce control: Control activities do not normally add value to customers. They are often designed to prevent problems from happening as a result of human mistakes. In several cases, however, control itself fosters neglect, with a negative impact on productivity. For example, a worker may not be careful enough when performing a business process activity because the worker knows that there will be some kind of control to catch mistakes. Additionally, some types of control, such as those aimed at preventing fraud, may prove to be more costly than no control at all. Some car insurance companies, for example, have found out that the cost of accident inspections for a large group of their customers, is much higher than the average cost of frauds that that group committed.
- Reduce the number of contact points: Contact points can be defined as points where there is interaction between two or more people, both within the business process and outside. This involves contacts between functions, and between functions and customers. Contact points generate delays and inconsistencies and, when in excess, lead to customer perplexity and dissatisfaction. In self-service restaurants and warehouses, for example, the points of contact are successfully reduced to a minimum. Additionally, it is much easier to monitor customer perceptions in situations where there is a small number of contact points. This makes it easier to improve business process quality.
- Execute activities concurrently: Activities are often executed in sequence, even when they could be done concurrently. This has a negative impact primarily on productivity, and is easier to spot on business process flowcharts than DFDs. In a car assembly process, for example, the doors

and other body parts can be assembled concurrently with some engine parts. This has been noted by several automakers, which, by redesigning their business processes accordingly, significantly speeded up the assembly of certain car models.

- Group interrelated activities: Closely interrelated activities should be grouped in time and space. Activities that use the same resources (i.e., artifacts or functions) may be carried out at the same location and, in some cases, at the same time. I often illustrate this point using the case of a telephone company that repaired external and internal house telephone connections (Kock, 1995a). This company had two teams, one team for internal and another for external repairs. An internal repair occurs, by definition, within the boundaries of a commercial building or residence; external repairs involve problems outside these boundaries. Whenever the telephone company received a customer complaint, it used to send first its internal team. Should this team find no internal connection problem, the external team would then be dispatched to check the problem. It took a business process improvement group to show the company that it was wasting hundreds of thousands of dollars a year, and upsetting customers due to repair delays, by not combining the two teams into a single repair team. This was because, when complaints were categorized and counted, it was found out that most of the problems were external.
- Break complex business processes into simpler ones: Complex business processes with dozens (hundreds in some cases) of activities and decision points should be broken down into simpler business processes. It often is much simpler to train people to execute several simple business processes than one complex business process. It is also easier to avoid mistakes in this way, as simple business processes are easy to understand and coordinate. In support of this point, I discuss in a previous publication (Kock, 1995a) the case of an international events organizer, which was structured around two main business processes—organization of national and international events. After a detailed analysis of these two business processes, which embodied over a hundred activities each, it was found that they both could be split into three simpler sub-processes—organization of exhibitions, conferences, and exhibitors' participation. This simplification improved the learning curve for the business processes, as well as reduced the occurrence of mistakes. It did not, however, lead to an increase in the number of employees needed. The reason is because, with simpler business processes, one person could perform functions in various business processes at the same time.

I now would like to make one important comment about this activity (i.e., the MetaProi activity called "search for suitable changes"). Here, business process improvement group members should not be so concerned about the feasibility of their redesign proposals. This concern will only limit the innovativeness of the redesign and, therefore, its effectiveness. Redesign feasibility analysis will be carried out at a later point, in an activity included especially for this purpose.

Redesign Stage: Incorporate Changes into the Business Process

In this activity, the facilitator should incorporate the changes proposed by the group into the business process flowchart or DFD and respective written description. This new business process model works as a feedback to the group, so that the proposed changes can be refined.

Guideline

The facilitator should try to state at this point who would be responsible for implementing the proposed changes in the business process. If such changes need involvement from higher management levels, this should be clearly stated. Such involvement may be needed, for example, for investment approvals and certain changes in the organizational structure.

Redesign Stage: Evaluate Redesign Feasibility

This is the last activity of MetaProi. In this activity, the group members should discuss the feasibility of the changes proposed to the business process so far and, if necessary, modify them to adapt those changes to the reality of the organization.

Subsequent Stages: Implement and Refine Redesign

The next stages are the initial implementation of the changes and their refinement, so they can be used in a routine way and, perhaps, in other similar organizational processes. The group can proceed on its own to these stages, provided that no involvement from higher management levels is necessary to implement the changes. If enough authority to approve and support the changes proposed can be found within the group, for example, and there are resources to carry this implementation out, then the group can proceed to business process change implementation right away.

If the above is not the case, the group should submit the change proposal to those who are in a position to have it implemented. Ideally, this should be done through the Business Process Improvement Committee, which is the committee responsible for the evaluation of redesign proposals and coordination of their implementation.

MetaProi in Practice: A College Example

The following example shows parts of a discussion carried out by a business process improvement group while attempting to improve a business process. The discussion takes place in a fictitious college, and the business process tackled relates to a practical introductory course in information systems. The course's name is Introduction to Business Computing; course code 0127A. The course had two conceptual classes per week, as well as 10 laboratory practical sessions all together. This discussion is based on a real discussion that took place in a similar context. Names and situations have been changed to protect confidentiality.

The discussion shown here was performed with the use of an e-collaboration system, which was led and facilitated by a person referred to here as Angus. Angus was also the graduate assistant of the introductory course that was targeted by the business process improvement group. Other group members were John, Anne, and Mark, who were faculty members involved in the design and teaching of course 0127A; Phil and Linda, who were members of the college's computer support staff; and Paula, who was director of the college's Computer Support Division, as well as Phil and Linda's boss.

Definition Stage: Discussing Problems

Angus selected and invited the six college members to join the business process improvement group. He did so based on several problems he had identified during course tutorials. Group members were invited via telephone, e-mail, and face-to-face conversations. They were given a description of the topic of the discussion and the group process to be followed—MetaProi. All those invited agreed to participate in the business process improvement group. The discussion then was started, as proposed in MetaProi, with a message sent by Angus to all business process improvement group members, describing the main problems observed during course tutorials. In the same message, Angus pointed out the main causes for the problems and asked the group members for comments. This message is shown here.

From: *Angus*

To: *John, Anne, Mark, Phil, Linda, Paula*

Date: *6 June 1995 1:04pm*

Subject: *0127A Course Problems (please give your feedback)*

In the first semester several problems contributed to lower the quality of the 0127A course (Introduction to Business Computing). In this message I try to identify the main problems found and what might have caused them. I might have forgotten something or mistaken certain causes, so please send your comments to either me or all the recipients of this message. Any contributions will be appreciated, including criticism.

The goal of this discussion is not to find someone to blame. It is to improve the quality of the course.

List of main problems observed in the first semester:

Students seemed to be, in general, upset by the workload. They complained that the course had too many assignments and that they were too time consuming.

From prac 5 on, the students didn't seem to understand what they were doing. Several students reported that these pracs were too "mechanical".

There are several minor mistakes in the manual, for all the pracs (annotated by me).

The sample files created for pracs 9 & 10 (using the software MS Access) don't fit on most of the computer HDs at the labs. Most of these computers have less than 2 Mb available on C: drive. The sample files created are above 1.5 Mb. This leaves very little working space for the operating system, which in turn starts to issue strange error messages.

Meetingworks (prac 4) didn't work properly, even after more than 7 previous tests where problems were found and supposedly solved.

Trader (prac 8) doesn't load sample files in the network. If you create the files on one machine, these files cannot be used on another machine. Platinum Software and Computer Support Division were contacted, but no solution was found so far. Platinum software says Trader was not designed to run on a network.

MS Access (pracs 9 & 10) was not configured to be used on the network. Certain functions (e.g., Wizards and print definition) cannot be used by more than one student at the same time.

Several hardware problems in the lab, such as: mice that don't work, print queues that don't match actual printers, machines that are too slow, and (mainly) print server breakdowns.

Possible causes for the problems:

- ***The way pracs were designed.*** *Again, I'm not trying to blame anybody. I think that the redesign of the pracs, keeping the problems above in mind, is one way to improve 0127A course quality.*
- ***The way communication is carried out today, between Computer Support Division and the lecturers.*** *For example, some pracs were designed assuming that the HDs in the lab had enough capacity. As there always are constraints in a process (e.g., resources that are likely to create problems) their identification is very important so further problems can be avoided. The way communication has been done so far does not foster the identification of those constraints.*
- ***The way the Computer Support Division handles requests.*** *I don't have knowledge of workload and how internal activities in the Computer Support Division are carried out, but I feel the consequences (e.g., requests take too long to be fulfilled, and recurrence of failures). Some of these consequences may well be caused by external causes (e.g., unbalance between resources and demand). A discussion of this point may require a separate thread, since I and*

the lecturers in this improvement group are not the only Computer Support Division customers.

Please give your feedback focusing on the problems and possible causes described above ASAP. As it is important that we concentrate our improvement efforts in one of the causes only, I'd like to ask you to choose one cause to be tackled by this group. This cause can be chosen from the list above or another suggested by a group member.

Angus

Angus, in this first message, set the main lines of the discussion comprising the first three activities of MetaProi (i.e., "identify problems," "identify business processes," and "select a business process for redesign"), which are the activities of the first stage of MetaProi, the definition stage. Below are a few of the replies from group members.

From: *Anne*

To:*Angus, John, Mark, Phil, Linda, Paula*

Date: *7 June 1995 12:20pm*

Subject: *0127A Course Problems (please give your feedback) Reply*

Interesting to hear about your experiences; I do a lot of my teaching in the lab. The general thrust of your comments matches my experience no matter how hard you try to set up modules, assignments etc to run 100% smoothly, there always seems to be problems.

Some of the problems are avoidable, some less so. Possibly your database assignment can be done with a database a lot smaller than 1.5mb? There does seem to be an inordinate number of operational problems in the lab mice not working, won't read disk drives etc at any time at least half a dozen PCs (in part) seem to be u/s for one reason or another.

Lotus has run very solidly for my courses in the last Semester WP6 has been a bit wobbly at times; MS Access collapsed for everyone on a couple of occasions but since then an update has been installed and it has been behaving. I don't use Trader but getting the network/ standalone business sorted out is obviously important.

I find the Computer Support Division to be pretty good and getting better. One of my problems is that my classes are 5 7pm on two nights, so if technical problems arise in the lab, its difficult for me to get rescued.

Despite the frustrations for both teacher and students, I feel that the overall value in raising students' computer literacy more than offsets this

Hope these comments help

Anne

From: *Mark*

To:*Angus, John, Anne, Phil, Linda, Paula*

Date: *7 June 1995 1:48pm*

Subject: *0127A Course Problems (please give your feedback) Reply*

1. *I don't think the assignments were too much. Perhaps the pracs were too time consuming. We need to reduce the tasks. The tasks were mechanical on purpose to facilitate marking. I don't see another way considering the resources we have.*
2. *Hardware /software problems. We just need to fix them. There are bound to be some hiccups with a course of this nature.*

Mark

From: *John*

To:*Angus, Anne, Mark, Phil, Linda, Paula*

Date: *8 June 1995 11:39am*

Subject: *0127A Course Problems (please give your feedback) Reply - Reply*

Anne's comments match my own experience but I don't think we are going to get anywhere trying to solve technical problems. Those are the Computer Support Division's responsibility. I think we should concentrate, as a group, on the way pracs are designed and run.

John

From: *Paula*

To:*Angus, John, Anne, Mark, Phil, Linda*

Date: *9 June 1995 12:18pm*

Subject: *0127A Course Problems (please give your feedback) Reply - Reply*

I agree with John's comments and think that we could discuss some of the problems initially listed by Angus in another group at Computer Support Division. I recognize there are problems in our Division and am open to suggestions of improvement.

I think that in this group, though, we should concentrate on the way pracs are run.

Paula

Analysis Stage: Summarizing Information

The discussion summarized through the above messages pointed towards the business process embodying the activities involved in designing and running the laboratory sessions (also referred to as "pracs") as the business process to be improved by the group. Based on this, the facilitator sent the following message to the business process improvement group.

From: *Angus*

To:*John, Anne, Mark, Phil, Linda, Paula*

Date: *12 June 1995 12:50pm*

Subject: *0127A Course Problems (an analysis of the process)*

Thank you all of you who contributed with comments on the problems found in the 211 course.

Anne and Mark surveyed student's perceptions both half way and at the end of the course. In this message I summarize the results of those surveys and add a flowchart model of the process of designing and coordinating 0127A pracs. Based on it and on the discussion so far, changes are proposed. Please add new changes or correct those that you think are inappropriate or will not work. I'll send you a third an final message within a few days with a list of the agreed changes. These changes will be then implemented and their results will be evaluated during the next semester.

Summary of surveys (perceptions of the pracs only):

- *The pracs provided practical and relevant grounding in using computers.*
- *There was too much work compared to other courses.*
- *There should be more tutorials and less pracs.*
- *There should be more lab times available.*
- *There should be more tutors in labs and prac marks should be returned faster.*
- *Prac labs were noisy and cramped.*

Flowchart model of the process of designing and coordinating pracs:

Please find the model attached. In order to read it, go to the option "view", choose the option "zoom" and then "zoom area". Then just select the area to be viewed in the flowchart[2].

Summary of proposed changes (and those who would be responsible for them):

In looking for activities in the flowchart that could be done at the same time or concurrently, I and John realized that a good deal of the markings could be done during the lab. This will probably reduce the amount of printouts per class and, consequently, the effect of print server breakdowns (Angus).

implify pracs in order to reduce workload (tutor & course lecturers).

1) *Move pracs 4 (Meetingworks), 9 and 10 (MS Access) to other labs outside the college, to a private course provider. or 2)have the problems with Meetingworks and MS Access sorted out here (Phil, Linda, and Paula).*

Figure 10.6. 0127A Course Processes Model

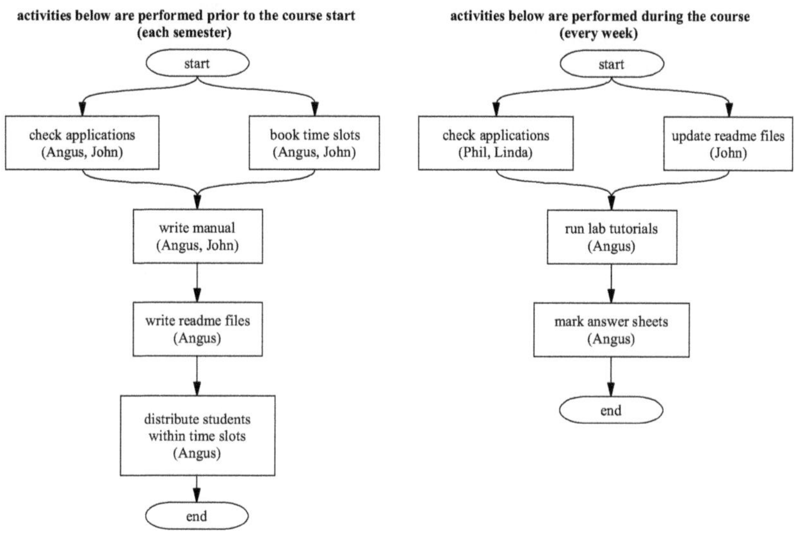

Change the exercises for prac 8 (Trader), so the students have to enter their own sample data to work with (John).

Check the applications used in the 0127A course in all the machines in the lab, every week, to make sure they are working properly (Phil and Linda). Angus would give in advance, to the Computer Support Division, a list of applications to be used in each week of the next semester .

Attach control cards to the machines so the students report problems and know the configuration of the machines and their previous history of problems (Paula).

The feedback from the members of this group to this message will define which changes will actually be put in practice. Send your input as soon as possible.

Thanks,

Angus

In his message, Angus, the facilitator, provides the background for three activities of MetaProi—"model the business process," "summarize performance information," and "highlight opportunities of improvement"—which are the activities of the second stage of MetaProi, the analysis stage. Below are a few of the replies from group members.

From: *Phil*

To: *Angus, John, Anne, Mark, Linda, Paula*

Date: *12 June 1995 2:41pm*

Subject: *0127A Course Problems (an analysis of the process) Reply*

Angus

a couple of comments on your mail message:

1. *on moving to use an outside lab several points*

   a. it will cost you

   b. i don't think you'll find someone with Meetingworks

   c. some of the MS Access problems we have are documented Microsoft problems of which they have no solution in a network environment...

2. *on checking apps being used.. it is already computer support policy that the labs are checked prior to a class if the course booking the lab communicates its intentions*

Phil

From: *Anne*

To: *Angus, John, Mark, Phil, Linda, Paula*

Date: *13 June 1995 5:09pm*

Subject: *0127A Course Problems (an analysis of the process) Reply*

Angus,

I think it will be hard for you to mark assignments in the lab, during the prac. You have your hands full helping these people. Marking is going to take a lot of extra time. I don't think you will have the time during lab hours.

Anne

From: *Angus*

To: *John, Anne, Mark, Phil, Linda, Paula*

Date: *14 June 1995 6:57pm*

Subject: *0127A Course Problems (an analysis of the process) Reply -Reply*

Anne,

In your message you say that I won't have time to mark assignments during the lab prac. That's a good point. However, in analyzing the process of running the pracs I realized that I get rushed off my feet mainly due to two reasons: 1)system problems and 2)complicated exercises. The main source of system problems are server problems, which seem to occur more often when the print queues are flooded with a large number of jobs (which is what normally happens when approx. 35 students try to print at the same time). In marking only those exercises that ask for printouts, hence waiving the respective printouts, I believe I may even reduce the rush during prac slots, since there will be less problems (not to mention the damaging effect of these problems to the image of the college). I'm not completely sure, though, this will happen.

The second reason, complicated exercises, is expected to have its impact mitigated with the simplification of the pracs.

Angus

Redesign Stage: Selecting and Evaluating Changes

The business process improvement group, after discussing the possible changes proposed by the facilitator and proposing new alternatives, clearly leaned towards some possible changes. The facilitator summarized these changes (see message below) sent to the members of the business process improvement group. In this message, the facilitator also asks the group about the completeness and appropriateness of the changes; that is, he asks the group for an evaluation of the redesign feasibility.

From: *Angus*

To:*John, Anne, Mark, Phil, Linda, Paula*

Date: *16 June 1995 4:56pm*

Subject: *0127 Course Problems (changes aimed at improving the pracs)*

Thank you all those who contributed with comments on how to improve 0127A course pracs.

Below is a summary of proposed changes based on our previous discussion. Please send your feedback on this summary to the discussion group (recipients of this message). Please let us know if you think some of the changes are inappropriate or the summary is not complete.

<u>Summary of changes to be implemented (together with an indication of those who are responsible for their implementation):</u>

- *Mark in lab, during the time slot, those exercises that ask for printouts. In these cases the printout will be waived when the student shows a suitable "print preview" (tutor).*
- *Reduce workload in pracs specifically at pracs 5, 6, 7, 9 and 10 (tutor and lecturers).*
- *Have MS Backup, or other backup program, easily accessible to the students so they can store large files (Computer Support Division).*
- *Solve Trader sample files problem (already implemented by Computer Support Division).*
- *Try to reduce the size of the sample files used at pracs 9 and 10. In case it's not possible, book another lab (the new lab that has been just built) for those pracs (tutor).*

Angus

In his message, Angus provides the background for three activities of MetaProi to be performed. These activities are "search for suitable changes," "incorporate changes into the business process model," and "evaluate redesign feasibility," which are the activities of the third stage of MetaProi, the redesign stage.

Note that Angus did not provide another graphical representation of the business process, since the changes in the business process did not lead to changes in the flowcharts. This was because the flowcharts did not represent the business process in much detail. The evaluation of the redesign feasibility was performed as the replies of the group members were discussed. Some of those replies are shown below.

From: *Linda*

To:*Angus, John, Anne, Mark, Phil, Paula*

Date: *17 June 1995 8:33pm*

Subject: *0127 Course Problems (changes aimed at improving the pracs) Reply*

Angus,

Given the context of you message I assume the purpose of backup is to store large files??...if this is the case there are better and easier (more reliable options you may wish to discuss this with me).

Linda

From: *Angus*

To: *John, Anne, Mark, Phil, Linda, Paula*

Date: *18 June 1995 5:07pm*

Subject: *211 Course Problems (changes aimed at improving the pracs) Reply*

Linda,

The purpose of backup is primarily to give students access to a tool for backing files up. Storing large files is a desirable spin off.

If you think there are better options, please choose the best among them and let me know how the students can use it. I've got no preference for MS Backup.

Angus

From: *John*

To:*Angus, Anne, Mark, Phil, Linda, Paula*

Date: *19 June 1995 5:07pm*

Subject: *211 Course Problems (changes aimed at improving the pracs) Reply*

I've already discussed some of the changes to reduce the workload in the pracs with Anne and Mark. I think it can be done without compromising the learning content. We are going to carry it out right away.

John

From: *Paula*

To:*Angus, John, Anne, Mark, Phil, Linda*

Date: *20 June 1995 5:07pm*

Subject: *211 Course Problems (changes aimed at improving the pracs) Reply*

I agree with the changes concerning the Computer Support Division. By the way, the problem with Trader, which has already been solved, as noted by Angus, was caused by the configuration of the network. Printing with Trader, though, is still impossible (at least to network printers). An alternative is to print to a file (most of Trader options allow it) and use Notepad or Wordperfect to print the file to a network printer - these programs work well on the network.

Paula

This marked the end of the business process improvement group discussion. The discussion shown here was summarized (i.e., not all electronic postings were shown) and targeted a simple set of problems. The business process improvement sought was a local one and involved business processes that would generally be seen as non-strategic. Authority to carry out the proposed changes and the respective resources needed were entirely found within the group. These changes were subsequently implemented and led to quality and productivity improvements that were seen as remarkable in the eyes of most group members.

Summary and Concluding Remarks

This chapter discussed MetaProi, a group methodology for business process improvement. MetaProi comprises a set of activities, guidelines, criteria, and graphical tools to be used by business process improvement groups. It assumes that groups will have from three to 25 participants who perform three main roles: (a) group leader, (b) facilitator, and (c) ordinary member. One person can play more than one role in the group (i.e., one person can be the group leader, the facilitator, and provide inputs as an ordinary group member). The group's main goal is to identify an organizational process in which improvement opportunities exist, and propose changes in order to translate those opportunities into practical improvement.

MetaProi is made up of three main stages—business process definition, analysis, and redesign. Each stage comprises the following interrelated activities.

- Business process definition stage
  - Identify problems
  - Identify business processes
  - Select a business process for redesign
- Business process analysis stage
  - Model the business process
  - Raise performance information
  - Highlight opportunities for improvement

- Business process redesign stage
  - Search for suitable changes
  - Incorporate changes to the business process model
  - Evaluate redesign feasibility

Each activity incorporates criteria, guidelines, and/or graphical tools for its effective completion. This chapter lists and discusses each of those elements, which are grouped by activity. It also shows how many of those elements can be put into practice through an example of a successful business process improvement group.

Endnotes

1 IDEF0 is the activity-based graphical modeling tool of the IDEF modeling standard. IDEF, in turn, stands for Integrated Computer Aided Manufacturing Definition. One of the reasons IDEF0 has been widely used in process modeling has been its early adoption as a standard process-modeling tool by the US Department of Defense (DOD) and other departments of the US government.

2 These instructions were designed for users of Groupwise, which was the e-collaboration system used to support the group discussion. The flow-chart model is provided in Figure 10.4 and contains two, rather than one, closely related business processes.

Chapter XI

A Close Look at Twelve Business Process Improvement Groups

A Structured Description of Several Groups

In this chapter, I provide a structured description of 12 business process improvement groups conducted at MAF Quality Management and Waikato University, both in New Zealand. I facilitated these groups based on the MetaProi methodology described earlier in this book, and the majority of the communication in these groups took place through an e-collaboration system. The group descriptions provided in this chapter and other group-related information have been used in several analyses discussed in previous chapters. Each group description comprises the following elements:

(1) Motivation: Describes why the business process improvement group was conducted. Here, I describe general problems faced by the organizations that called for business process improvement attempts. I also describe related pressures faced by prospective group leaders, which motivated them to undertake the business process improvement initiatives.

(2) Formation: Includes how the business process improvement group was begun and identifies the individuals invited to take part in the business process improvement group discussion.

(3) Features: Includes general features of the business process improvement group, such as duration, total number of members, number of members who posted at least one electronic message to the group, number of electronic postings exchanged by members, and proportions of total time spent by members interacting electronically and orally.

(4) Stages: Describes how the business process improvement group stages of business process definition, analysis, and redesign were conducted.

(5) Results: Describes the outcomes of the business process improvement group discussion in terms of actual business process improvement and short-term organizational implications.

As mentioned before, the business process improvement groups described in this chapter were taken from two New Zealand organizations. Nevertheless, I believe that the range of situations that are reflected in those groups is fairly broad, and the behavior of the group members is typical of what one would probably encounter in other countries, especially Western Hemisphere countries. This opinion is based on my past experience facilitating over 100 business process improvement groups in the U.S. and other countries. In my view, the range of situations addressed in this chapter is broad enough so that readers can have a good general idea of what e-collaboration technology-supported business process improvement groups are all about.

As discussed in previous chapters, there are striking similarities between the e-collaboration technology-supported business process improvement groups I studied in New Zealand and other similar groups I studied in the U.S. and other countries. Those similarities allow me to conclude that the range of situations covered in this chapter is fairly comprehensive.

The e-collaboration system used by the business process improvement groups described in this chapter was developed based on the Novell GroupWise e-collaboration environment, commercialized by Novell Corporation. Essentially, the e-collaboration system enabled the creation of one e-mail distribution list for each business process improvement group, as well as access by group members to discussion archives and other business process improvement-related documents.

MAF.G1: Software Support

Motivation

MAF Quality Management's regional information technology support department was established to provide general information technology support to MAF Quality Management's offices located in New Zealand's North Island. Such support typically involved the solution of hardware and software problems upon request from the users, and preventive upgrades in the region's information technology infrastructure, including local connections and local area network operating system software.

Recently, the department had been receiving a number of complaints from users, particularly relating to the departmental business process of providing software support for users. Three main problems, from the users' point of view, appeared to be at the source of those complaints:

- A perceived slow turnaround of software repair jobs
- A lack of information about the status of their jobs (i.e., requests for software application repairs)
- A high number of problems related to their local area networks, seen as likely caused by the department's limited time to carry out preventive local area network software maintenance

Formation

One of the department's team leaders decided to lead this group (referred to here as MAF.G1) and invited four other employees of the same department to take part in the discussion. Two of the invited members worked in the same office as the group leader. Their organizational seniority was slightly lower than the group leader's. One of the other members was the acting regional manager of the information technology support department, the group leader's boss, whose office was about 100 meters down the hall in the same building as the group leader's. The fourth prospective member was based in a different city and was subordinate to the group leader. All those invited agreed to participate in the discussion.

Features

The group lasted 28 days and was comprised of five members from only one department, but it was based in two different cities. From the five individuals who agreed to participate, only three members contributed postings to the discussion. The interaction in this group comprised seven electronic postings, and a number of one-on-one phone and face-to-face conversations.

According to estimates provided by group members, 83% of the total time spent by group members in the group discussion was in interactions through the e-collaboration system, whereas the remaining 13% was in oral one-on-one interactions.

Stages

In the definition stage, the group leader posted a message to the group containing a list with three problems and two business processes, suggested as the main causes of the problems and, thus, as the target for the redesign. This posting was followed by two replies with comments on the leader's message. These replies led to the selection of one business process for redesign.

In the analysis stage, the leader posted a message to the group containing a textual description of the selected business process, a summary with performance-related information, and eight proposed changes. This posting was followed by two replies refining the eight business process changes proposed.

In the redesign stage, the group leader posted a message to the group with a description of the eight refined business process changes agreed on by the group, the names of those responsible for implementing the changes, and the change implementation deadlines. No replies followed this posting.

Six out of eight business process changes proposed by the group were implemented within six months of the group's completion.

Results

A qualitative evaluation of the changes, based on departmental staff perceptions, indicated a slight increase in the perceived quality and productivity of the redesigned business process, and a partial solution of the three main problems used as a starting point for the business process improvement group discussion.

The group leader perceived the outcome of the group discussion as positive and noted that the group discussion had allowed the department to tackle a number of problems that had been long awaiting solutions. For a number reasons, including busy individual timetables, these problems had not been dealt with for over six months prior to this business process improvement group.

MAF.G2: The Internal Newsletter

Motivation

By the end of the facilitation of group MAF.G1, I was invited by the editor of MAF Quality Management's internal newsletter to facilitate a business process improvement group expected to deal with three main problems:

- The lack of contributions to the newsletter
- Recurring distortions, according to a number of newsletter contributor complaints, in the articles contributed by employees when they were finally published in the newsletter
- Delays in the publication of several contributed articles, which in the past had led to the publication of outdated information

All these problems were perceived by the editor as being linked to the process of editing the newsletter. More specifically, the editor believed that the communication process between contributors and the newsletter's editor was at the source of most of those problems.

Formation

The newsletter's editor decided to lead the group. He invited four other members to take part in the group discussion. All of the invited members worked for the same division, a communications support division, and were based in different cities. Their day-to-day work was in different departments; they were involved only part-time in the business process of editing the internal newsletter. Of the prospective group members, three, including the group leader, were in senior management or technical positions, and the other two were in assistant positions. All of the invited members agreed to participate in the business process improvement group discussion.

Features

The group lasted 26 days and, as already indicated, was comprised of five members from four departments in the same division. The members were based in four different cities. From the five agreeing to participate, only four members contributed postings to the discussion. The interaction in this business process improvement group comprised nine electronic postings, and a number of one-on-one phone conversations.

According to estimates provided by group members, 89% of the total time spent by group members in the group discussion was in interactions through the e-collaboration system, whereas the remaining 11% was in oral one-on-one interactions.

Stages

In the definition stage, the group leader posted a message to the group containing a list with three problems and two business processes that suggested the main causes of the problems and, thus, the target for the redesign. This posting was followed by three replies with comments on the leader's message, whose content led to the selection of one business process for redesign.

In the analysis stage, the leader posted a message to the group containing a textual description of the selected business process split into eight activities, a summary with performance-related information, and three proposed changes. This posting was followed by one reply refining the business process changes proposed.

In the redesign stage, the group leader posted a message to the group with a description of the three refined business process changes agreed on by the group and the names of those responsible for implementing the changes right away. Two replies followed this posting agreeing with changes and praising the initiative.

All business process changes proposed by the group were implemented within three months of the group's completion.

Results

The newsletter editor perceived the group discussion as extremely useful, and said that he intended to conduct other similar business process improvement groups in the future in his area. The editor reported the problems that originated with the group as having been completely eliminated as a result of the business process changes.

MAF.G3: Pest and Disease Outbreaks

Motivation

One of the key business processes performed by MAF Quality Management was media liaison during pest and disease outbreaks threatening New Zealand's agriculture. Such outbreaks had typically occurred no more than three times a year prior to this business process improvement group.

During several outbreaks in the past, a main problem had been a lack of unified purpose and work coordination of the several teams reporting outbreaks to the media (mostly TV and radio stations, and newspapers). This had previously led to the release of misleading information to the public, as well as information that was not consistent for different parts of the country. Dealing with this problem was now a responsibility of a senior manager in the communications support division who had been a member of group MAF.G2.

Formation

The senior manager, motivated by his experience in group MAF.G2, requested me to facilitate a business process improvement group led by him. He invited six other members to take part in the group discussion, all from the communications support division. Four of the invited members were senior managers or consultants; the other two were assistant consultants. The main background of all involved was, homogeneously, business communication. The invited members worked in offices located across five different cities; only two of them, a senior and an assistant consultant, were based in the same office and city. All of the invited members agreed to participate in the discussion.

Features

The group lasted 14 days and, as already discussed, was comprised of seven members from one department and six different cities. From the seven agreeing to participate, only three members contributed postings to the discussion. The interaction in this group comprised four electronic postings, a number of one-on-one phone and face-to-face conversations, and one group face-to-face meeting.

According to estimates provided by group members, 18% of the total time spent by group members in the group discussion was in interactions through the e-collaboration system, whereas the remaining 82% was in oral one-on-one or group interactions.

Stages

In the definition stage, the group leader posted a message to the group containing a list with three problems and two business processes that suggested the main causes of the problems and, thus, the target for the redesign. This posting was followed by two replies with comments on the leader's message. This led to the selection of the two business processes initially proposed by the leader for redesign.

The analysis stage was conducted through a one-day group meeting chaired by the leader, where seven business process changes were discussed and refined by the group.

The redesign stage was conducted in the week after this meeting, where the group leader posted a message to the group with a description of the seven refined business process changes agreed on by the group, and the names of those responsible for implementing the changes right away. There were no replies to this posting. Four out of the seven proposed business process changes were implemented within four months of the group's completion.

The group leader, who was responsible for the coordination of the implementation of most of the business process redesign proposals, left the organization before the next outbreak occurred.

Results

According to members of the communications support division, the business process changes led to a moderate increase in the quality and a reduction in the cost of the target business processes. These improvements were reported when the communication team was brought into action again during a fruit-fly outbreak approximately seven months after the business process improvement group was completed.

MAF.G4: Quality Management Consulting

Motivation

Right after group MAF.G3 was completed, I was introduced by MAF.G3's leader to a middle manager who had been assigned to solve a corporate-level problem. MAF Quality Management had a line of consulting products (i.e., consulting service products) targeted at the implementation of quality systems based on quality standards set by the government for the New Zealand food and plant industries. In addition, some recent standard-related changes called for a new consulting product that not only would address those changes, but also would be well integrated with the existing products.

MAF Quality Management's consulting products were tied to quality standards whose accreditation was the responsibility of a national accreditation body that had recently launched a new standard aimed at small businesses. As a result of this launch, MAF Quality Management had to quickly develop a new consulting product tied to this new standard and, consequently, redesign some other consulting products to ensure product synergy. At the same time, MAF Quality Management had to avoid overlaps between this new product and other products already established in the market.

Formation

After a few meetings with me, in which the group methodology was discussed, the middle manager decided to initiate a business process improvement group discussion. He invited eight prospective group members. The invited members were from four different departments and based in five different cities. Two of the invited members were senior managers; the other six had middle management or consulting positions.

Two main business backgrounds were evenly represented in the group, as related to the main business areas at MAF Quality Management, namely "food" (mostly meat and dairy) and "plant." Unstructured interviews with employees prior to this group discussion suggested that each background was closely related to a specific business culture, food personnel being seen (by most of MAF Quality Management's employees with whom I talked) as more aggressive and competitive. Food personnel were also seen as less polite than those who worked in the plant business, who were in turn typically seen as calm, cooperative, and polite.

All of the invited members agreed to participate in the discussion. Two other senior managers were included on request right after the business process improvement discussion began.

The group leader decided to conduct the discussion in a less structured way than proposed by MetaProi, starting the group stages with more "subjective" postings and giving opportunity for members to, in his words, "define the actual focus of the group."

I expressed concern about this lack of objectivity in the group discussion and the likely negative consequences, such as a possible lack of group focus. However, since I agreed to facilitate the group, whatever the leader's decision might be, the group leader decided to carry on with his devised strategy at his own risk.

Features

The group lasted 30 days and, as already discussed above, was comprised of 11 members from five departments and six different cities. From the 11 agreeing to participate, only seven members contributed postings to the discussion. The interaction in this group comprised 18 electronic postings and a number of one-on-one phone and face-to-face conversations.

According to estimates provided by group members, 80% of the total time spent by group members in the group discussion was in interactions through the e-collaboration system, whereas the remaining 20% was in oral one-on-one interactions.

Stages

In the definition stage, the group leader posted a message to the group defining the main problem to be tackled by the group in a very general and abstract way, and asking members to reply with the points on which they thought the group should focus. This posting was followed by nine replies with general and subjective comments on the main problem described; new issues were raised, some totally unrelated to the main problem posed to the group.

Noticing that the group discussion was lacking focus, the group leader posted a message summarizing previous postings and providing a more focused list of problems, related business processes, and business process changes. This posting was followed by six unfocused replies; one from a senior manager angrily criticized the group's indecision.

Forecasting a likely lack of agreement and perhaps conflict in the group, the leader decided to discontinue the discussion. He did so by means of a wrap-up posting calling for a face-to-face meeting to discuss whether the new quality standard should be adopted, and if so, which changes in other consulting products should be implemented.

Results

No business process changes were proposed. The group leader saw the group discussion as a failure. A month later, the same set of issues tackled by this group was discussed in a different face-to-face group meeting, but, again, no agreement was achieved.

MAF.G5: Information Technology Support

Motivation

This group grew out of the need to target the business process for providing information technology support to internal users as a whole, rather than only software users, as in group MAF.G1. In a sense, this group discussion was an extension of the one conducted by group MAF.G1.

Here, however, more general problems facing the information technology support team were targeted. Two other differences were that this group involved fewer information technology support employees, with most of its members being business process customers; in addition, the person interested in leading this group was not a manager, but a senior computer support person. This person contacted me and asked me to facilitate the group discussion while group MAF.G4 was still being conducted.

Formation

The group leader invited 14 members to take part in the group discussion. Four of these members worked in the information technology department. One of them was the acting regional manager. The other 10 were business process customers in senior and middle management, as well as technical non-management positions. The invited members were from six different departments based in offices located in nine different cities. All those invited agreed to participate in the discussion.

Features

The group lasted 26 days and was comprised of 15 members from seven different departments spread throughout 10 different cities. From the 15 agreeing to participate, only 11 members contributed postings to the discussion. The interaction in this group comprised 23 electronic postings and a number of one-on-one phone and face-to-face conversations.

According to estimates provided by group members, 77% of the total time spent by group members in the group discussion was in interactions through the e-collaboration system, whereas the remaining 23% was in oral one-on-one interactions.

Stages

In the definition stage, the group leader posted a message to the group containing a list with four problems and five business processes suggested as the main causes of the problems and, thus, as the target for the redesign. This posting was followed by nine replies emphasizing some of the problems and proposing possible solutions. The analysis of these replies indicated an emphasis on two of the five business processes initially discussed.

In the analysis stage, the leader posted a message to the group containing a textual description of the two selected business processes, a summary with performance-related information, and a request for change proposals. This posting was followed by six replies proposing business process changes, from which five focused business process changes were identified.

In the redesign stage, the group leader posted a message to the group with a description of the five business process changes proposed by the group, the names of those responsible for implementing the changes, and implementation deadlines. Five replies followed this posting, refining the changes and praising the initiative.

Four of the five business process changes were implemented within six months of the group's completion.

Results

A survey of customer perceptions indicated an increase in perceived business process quality. Perceptions of those who carried out activities in the business process suggested an increase in the productivity of the business process. During the implementation of the business process changes, the leader of the information technology team was promoted to the position of regional manager of information technology support.

MAF.G6: Employee Training and Development

Motivation

This group arose from the need to improve the productivity and quality of the business process of coordinating employee training and development. MAF Quality Management's large employee base had to be centrally and constantly monitored to ensure that employees would possess the skills and the knowledge background needed in the projects they were routinely assigned by their managers. These skills and knowledge background changed constantly. Among the change drivers were the introduction of new regulations or quality standards in the market, as illustrated by our description of group MAF.G4.

Monitoring and coordinating workers' qualifications and informing managers of the need for worker training was the responsibility of a multi-disciplinary team, which we shall call here "staff training coordination team." This team was comprised of members from several departments, and was facing a number of problems at the time that I made contact with one of its members. Among the most pressing problems was the lack of a flexible and ease-to-use classification scheme for the description of employee qualifications embodying MAF Quality Management's constantly changing product portfolio.

Formation

During a chat with managers in the tea room, where I aired the intention to approach a possible leader for a sixth business process improvement group to be facilitated at MAF Quality Management, I was told of some problems being faced by the staff training coordination team. Subsequently, I approached one of the senior members of this team and offered my services as a group facilitator.

After some meetings at which the senior team member was briefed about MetaProi and the results of previous groups, he decided to lead a business process improvement group and invited 13 other members to take part in the group discussion. The invited members were from two different departments and were based in seven different cities. They worked in different but cooperating departments, and were involved only part-time in the business process of coordinating employee training and development. Of the prospective group members, about half were in senior management or technical positions; the other half held assistant positions.

Assuming that most group members previously agreed that the key point to improving the target business process was to develop a computer system, the prospective leader decided to use the group to define computer system requirements. All of the invited members agreed to participate in the discussion.

Features

The group lasted about 10 days as a business process improvement group and was subsequently established as a semi-permanent forum for a discussion about software requirement specifications for approximately another three months. The group was comprised of 14 members from three departments and eight different cities. From the 14 agreeing to participate, only six members contributed postings to the discussion. The interaction in this group comprised six electronic postings and a number of one-on-one phone and face-to-face conversations.

According to estimates provided by group members, 67% of the total time spent by group members in the group discussion was in interactions through the e-collaboration system, whereas the remaining 33% was in oral one-on-one interactions.

Stages

As mentioned earlier, the group leader started the group with the assumption that group members agreed that the key point to improve the already defined target business process was to develop a computer system. Therefore, in the first message, the leader tried to steer the group members into specifying the requirements of the computer system.

The group members reacted to this attempt by posting replies that were perceived by the group leader as "trivial," as well as suggesting lack of enthusiasm about the group discussion. After several phone and face-to-face contacts with group members, in which they were asked to contribute more and better structured information, the leader grew increasingly frustrated and finally decided to use the group discussion as a forum for permanent exchange of information about software requirements, rather than for business process improvement. The goal of this forum was to complement face-to-face meetings.

Results

The group leader saw this group as a failure from a business process improvement perspective. Moreover, three months after the business process improvement group was discontinued and the e-collaboration system was defined as a forum to complement face-to-face computer system requirement specification meetings, I checked the e-collaboration system and noticed that there had been no electronic contributions from group members.

However, a further conversation with a member of the staff training coordination team suggested that the computer system requirement specification work was still underway, being most of the interaction carried out in face-to-face group meetings and one-on-one electronic messages between team members (these were not captured by the e-collaboration system, which only archived electronic postings to the whole group).

Waikato.G1: A Practical Computing Course

Motivation

Waikato had recently set up a successful theoretical and practical computing course aimed at building student business computing skills on a number of software applications, including e-mail, group decision support systems, Internet Web browsers, word processors, spreadsheets, and database management systems.

While successful, with about 100 enrollments per semester, the course recently had been the focus of an avalanche of student complaints, which were perceived by management as negatively affecting Waikato's image. Most of the complaints were linked to the frequent computer problems that were experienced by students when trying to run their practical course assignments.

Formation

The person most directly involved with the practical component of the course was the course tutor, who was responsible for practical demonstrations in a computer laboratory during the time slots assigned for the course.

Students and management both had urged the course tutor to come up with innovative solutions to reduce or eliminate computer problems experienced by students. His first step was to bring together members of two departments—

a computer support and an academic department—that were directly involved in the practical component of the course.

The members initially were reluctant to meet and discuss those problems because of previous conflicts between the two departments, who blamed each other for the problems in the course. However, they eventually agreed to participate in the business process improvement group led by the course tutor when they knew that the discussion would take place through an e-collaboration system. E-collaboration technology mediation was perceived by most of those prospective members as likely to lead to a less formal and less confrontational discussion, different from what, in their view, would likely happen in face-to-face meetings.

Features

The business process improvement group lasted 33 days and was comprised of seven members from the computer support and the academic departments. The interaction in the group comprised 21 postings, and a number of one-on-one phone and face-to-face conversations.

According to estimates provided by group members, 71% of the total time spent by group members in the group discussion was in interactions through the e-collaboration system, whereas the remaining 29% was in oral one-on-one interactions.

Stages

In the definition stage, the group leader posted a message to the group containing a list with eight problems and three business processes, which were suggested as the main causes of the problems and, thus, as the target for redesign. This posting was followed by four replies with comments on the leader's message, whose content led to the selection of two business processes for redesign.

In the analysis stage, the leader posted a message to the group containing a flow chart description of the two selected business processes (as a file attached to the message), a summary with performance-related information, and six proposed changes. This posting was followed by 13 replies, and the group agreed on five business process changes.

In the redesign stage, the group leader posted a message to the group with a description of the five business process changes agreed on by the group, the names of those responsible for implementing the changes, and change implementation deadlines.

Results

The group was completed in time for change implementation and impact assessment during the following course semester. All changes were implemented either before or during the course semester, and their impact was assessed through a survey of student perceptions about the course. The survey covered most of the points targeted by the business process improvement group, and its results indicated a remarkable improvement in the quality of the course when compared with the results of a survey performed in the previous semester.

Waikato.G2: Academic Advice for Students

Motivation

The general problem that this group wanted to discuss was the increasing level of litigation brought against Waikato by students, with regard to the academic and program advice they received from staff and faculty members. More specifically, this general problem included issues such as the following:

- Increased instances of litigation by students having made significant academic program decisions based on incorrect advice
- Increased instances of the head of the organization having to use his special (regulatory) powers to waive regulations due to litigation
- Faculty members having differing expectations of administrative staff with regard to who should give what advice to students

Formation

The self-appointed group leader was a senior administrative staff member that reported directly to the head of one of the Waikato's business units. This particular business unit was chosen because it was at the source of most of the student litigation problems in the previous year.

A number of people were invited and agreed to contribute to the business process improvement group. Two of them were intermediate administrative staff, and one was a senior administrative staff that had previously been involved in strategy-focused discussions about this issue. Two among the other invited people were faculty members, and, finally, two others were heads of academic departments.

Features

This business process improvement group lasted 41 days and involved five departments, represented by the eight members just mentioned. All of the eight members who agreed to participate in the group contributed postings to the discussion. The interaction comprised 30 postings and a small number of one-on-one telephone and face-to-face conversations.

According to estimates provided by the group members, 96% of the total discussion time spent by group members was in interactions through the e-collaboration system; 4% of the time was in oral one-on-one interactions.

Stages

In the definition stage of the discussion, the leader posted a message that detailed four problems and two potential business process causes. Initially, two causes were proposed: (a) that there was a lack of training on the part of key faculty and general staff members; and (b) that students did not have immediate access to faculty members and, therefore, should seek the advice of administrative staff members. Thirteen replies that followed this initial message commented on the group leader and others' postings, and proposed change suggestions focused on two business processes.

In the analysis stage, these two business processes were described, and two proposals for business process change were made. This message elicited five responses that modified the tentative change proposals. Unfortunately, at this stage, some information had been missing from the discussion because of the confidential nature of the litigation issue. However, one of the departmental heads had candidly contributed potential issues from his own department, which, in some way, eliminated that particular gap in the discussion material.

Finally, in the redesign stage, two changes were proposed, consolidating previous change proposals. This posting was followed by nine replies commenting on the changes and proposing minor amendments. While one of these change proposals received almost unanimous support from the group, the other was only partially supported.

Results

Although a timeline and responsibilities were set for the implementation of the two business process changes resulting from this group, these changes actually were never implemented. It was found later that one of the members assigned responsibility for implementing the business process changes was not really committed to implementing the changes. This person perceived the discussion, in her own words, as lacking "legitimacy and authority."

Waikato.G3: Student Computer Support

Motivation

Some time after Waikato.G2 was completed, and shortly before new computing facilities were to be made available, a discussion was initiated to improve the entire student computing experience at one of Waikato's computer facilities. The facility housed 150 computers and seven printers. Soon, a new area would be made available, adding 75 new machines.

Although the computer system and support staff were able to accommodate teaching requirements, providing computer support to students was generally seen as a troublesome task. Many problems were caused by changes to courses by academic departments that were not effectively communicated to computer support staff.

On the other hand, changes in hardware and software implemented by computer support employees in the facility, which were aimed at alleviating previous problems, ended up causing new problems that were not being communicated back to faculty members. Those faculty members needed this feedback in order to change course material. These problems were similar to those found in Waikato.G1.

Formation

A course tutor, who had also been a student at the institution, led the business process improvement discussion. He worked in the computer facility in a user support role, and was familiar with concepts and ideas related to business process improvement. The manager of the computer facility was eager to obtain some strategic input from outside of his immediate support staff. The leader convinced computer support staff and senior members of five teaching units who had computer related material in their course curricula, to become members of the business process improvement group.

Features

The business process improvement group lasted 32 days and involved 11 members from five departments. From the 11 agreeing to participate, only seven members contributed postings to the discussion. The interaction in the group comprised 23 postings, as well as some one-on-one oral (mostly face-to-face) conversations.

According to estimates provided by the group members, 81% of the total time spent by group members in the group discussion was in interactions through the e-collaboration system; the other 19% was spent in oral one-on-one interactions.

Stages

In the definition stage, the leader posted a message to the group containing a list with five problems phrased as generic areas of complaint by students, and four business processes to be targeted for redesign. This posting was followed by four replies developing two distinct themes. The first theme was that some faculty members considered complaints from the students as being outside of their area of responsibility. The second theme was that workable solutions did, indeed, exist, and several members offered a variety of these.

At this point, the leader tried to force the discussion onto the business processes proposed by him through a message posted to the group. This was done in an attempt to focus the discussion and also to create a climate for the development of appropriate solutions. This message, however, was largely ignored in the six replies that followed.

In spite of this, the leader proceeded to the analysis stage with a posting in which he described one of the business processes (split into 10 main activities) by means of a non-standard diagram attached to the posting, and a list with three proposed changes. This was followed by five replies, again, largely unrelated to the leader's posting.

A "harsh" posting from the leader aimed at redirecting the discussion was rebutted in a reply from one of the members, a senior lecturer. At this point, the leader, upon advice from the facilitator, posted a message to the group apologizing for his "harsh tone" in the previous posting. This was followed by a reply that, again, was largely unrelated to the main topic of the discussion.

A concluding message by the leader highlighted the major themes of the discussion and concluded that the computer support staff members were the best people to decide their own fate. No replies followed.

Results

Although the outcomes of this group were not very positive in business process improvement terms, since the group failed to generate any business process improvement proposal, group members generally were pleased with the discussion. Some members noted that the discussion promoted dissemination of interdepartmental knowledge, while others, notably computer support staff, saw the group discussion as an endorsement of their good work.

Waikato.G4: Student Assignments

Motivation

The amount of course-related data exchanged between faculty and students at Waikato has led, over time, to the creation of support departments whose common characteristic was to coordinate the handling of these data, particularly data going from instructors to students and vice-versa, in the form of handouts and completed assignments. Such departments were typically within a school of studies (e.g., school of management studies).

For the sake of this group's description, I shall generically refer to these support departments as student handout places (SHOPs). Course-related data flowing from teachers to students in SHOPs typically involved course assignment instructions, course readings, marked assignments, and documentation about external organizations; the latter typically used in courses with fieldwork content. Course-related data flowing from students to faculty in SHOPs typically involved completed assignment forms.

This business process improvement group emerged out of the interest of a SHOP manager, who had recently arrived at Waikato, in turning the business process of handling the data flowing within her SHOP, particularly student assignments, into a more efficient one.

Formation

I met the SHOP manager informally at a tearoom while talking to the leader of a previous group, and after being told about the manager's ideas about possible improvements in the SHOP, I offered to facilitate a business process improvement group led by the SHOP manager. She decided to invite three faculty members, two students, and one administrative staff to participate in the group discussion.

Features

The group lasted 45 days and involved seven members from five departments. From the seven members who agreed to participate, only five contributed postings to the discussion. The interaction in the group comprised 15 postings and some one-on-one oral (mostly face-to-face) conversations.

According to estimates provided by the group members, 77% of the total time spent by group members in the group discussion was in interactions through the e-collaboration system; the other 23% being in one-on-one oral interactions.

Stages

In the definition stage, the leader posted a message to the group containing a list with five problems and five business processes, presented as the possible sources of the problems. This posting was followed by nine replies focusing on two of the business processes listed. These business processes related to the actual handling of course material, and the way communication between faculty members and SHOP staff was conducted regarding course assignment handling issues.

In the analysis stage, the group leader posted a message to the group summarizing contributions and providing a textual description of two business processes. In this message, she also summarized some qualitative information about the performance of the business processes, and promised to look up some extra technical information to be summarized in a future posting. She also proposed four business process changes. Two replies followed this posting with general comments, one of them questioning the completeness and correctness of the business process descriptions.

Ignoring this criticism, the group leader then posted a message summarizing some technical issues related to the implementation of the business process changes, which pointed to the implementation of two changes relating the setting up of three items. The first was a new computerized bar-coding system for keeping track of course material. The second item related to Internet Web pages with standard cover sheets for the assignments. The third was in connection with bar codes for the students.

This was followed by one final wrap-up posting from the group leader summarizing two proposed business process changes, responsibilities and implementation deadlines. The main responsibility for the coordination of the implementation of the two business process changes lay on the group leader.

Both of the business process changes were partially implemented in pilot projects within three months of the group completion, and their full implementation was scheduled to be conducted within the next six months. One posting from the leader followed shortly after with some more details about business process change implementation issues. No replies followed.

Results

An assessment of the pilot projects, which involved faculty members, SHOP staff, and more than 150 students, indicated a drastic increase in business process productivity and a moderate increase in the quality of the business process. The computerized bar coding system, in particular, was seen by the SHOP's manager, as well as by some teachers, as likely to allow SHOP staff to move from predominantly manual and "mechanical" activities towards knowledge-intensive roles. Examples of the latter were providing support to students and academics as to where and how to find relevant information for research and courses.

Waikato.G5: International Graduate Students

Motivation

Waikato's support structure to international students comprised a central international student's office plus a number of decentralized departments within schools of studies. One major problem facing one of these decentralized departments, known at Waikato as international student departments, was that permanent resident students were not required to meet the same English standards as international students. Permanent resident students were those who had been born outside New Zealand and had legally migrated to the country, or who were children of permanent residents.

International students had to fulfill a number of requirements, including a minimum score in the Test of English as a Foreign Language (TOEFL) before they could enroll in any course. While the international students office was prepared to cater to the needs of these students, it did not have the resources to provide support for permanent resident students whose English language skills were far below those of entry-level international students. As permanent resident students received the same treatment as New Zealand nationals, they were not considered in budget allocation decisions regarding the international student department.

Nevertheless, a number of permanent resident students were routinely asked by their instructors to seek support from the international student department. In addition to an observable lack of English language skills, one of the other reasons for that was because, as one of the lecturers put it, "They [the permanent resident students] look like international students…." In fact, most international students in New Zealand, as well as immigrants in general, were from Asian countries. Thus, they were indiscriminately seen as members of the same group, as far as academic issues were concerned. Most lecturers, typically from European or North-American backgrounds, saw these students as simply "international students."

Formation

I was advised by one of Waikato's senior professors, who had been facing difficulties in dealing with permanent resident students, to contact the international students' department manager. When approached, this manager told me that she had asked for a budget increase so she also could cater to the needs of permanent resident students. She wanted to embrace this responsibility, however problematic it might me, because, in her words, "If we don't do it, the competition will…." Moreover, Waikato had recently tried to build up an image of a multicultural institution, and this was an opportunity for Waikato to reaffirm its commitment.

On the other hand, the manager admitted to having experienced some difficulties in the past with her superiors because of her ideas. Among other things, they seemed determined to postpone any decision regarding differential treatment to permanent resident students. She saw the business process improvement group as a last resort to resolve the permanent resident students issue, and decided to invite a broad range of members to take part in the group. Among these were some of her superiors and faculty members, as well as international and permanent resident students.

Features

The group lasted 33 days and involved 13 members from eight departments. From the 13 agreeing to participate, only eight members contributed postings to the discussion. Three out of the five members who contributed no electronic postings to the discussion actively participated in a face-to-face group discussion conducted in the analysis stage of the group. The interaction in the group comprised 22 postings, a number of one-on-one electronic messages, some one-on-one oral conversations, and one group face-to-face meeting.

According to estimates provided by the group members, 52% of the total time spent by group members in the group discussion was in interactions through the e-collaboration system; the other 48% was in oral interactions.

Stages

In the definition stage, the leader posted a message to the group containing a list with four problems and three business processes, presented as the possible sources of the problems. These business processes related the follwing:

- The way support was provided to students with English problems
- The way communication between the department and students was conducted
- The method by which students were matched with supervisors

This posting was followed by six replies, one of these correcting the description of some of the problems and expanding the scope of one of the business processes presented. The other five postings provided general and somewhat vague comments on some of the issues raised in the first message.

At this point, the group leader was summoned to a meeting with her superiors and was advised to discontinue the business process improvement group discussion. Although the manager refused to discuss the entire content of this meeting with me, it became apparent that her superiors saw the conduct of the discussion through the e-collaboration system as inappropriate, as the discussion could trigger comments about racial issues. They feared that discussion transcripts could later be used against Waikato in formal complaints and perhaps lawsuits by students from ethnic minorities.

As a result, the manager decided to shift the focus of the group discussion towards a related and less politically sensitive issue, namely the nature of the support provided to international graduate students. She tried to do this in a posting where she explained to the group that the change in the focus of the group was due to low response rate. She subsequently described five main problems. She also described one business process, which was presented as the source of the problems.

The reason why the manager decided not to tell the group the real reason for the shift in the focus of the group was that her superiors requested that their previous meeting be kept secret. This posting was followed by 12 replies. Only five of them were related to the new discussion topic. The remaining seven postings were related to the old topic; that is, permanent resident students support. These postings suggested some unease about the sudden change of topic.

Given the seeming perplexity of some group members, a face-to-face meeting then was called by the group leader to conduct the next stage of the group discussion. Six group members attended the face-to-face meeting, which lasted approximately two hours and 15 minutes. In this face-to-face meeting, the group leader stuck to the new topic. She supplied the group with a business process description and performance-related information, and proposed four business process changes for discussion.

The group agreed on three business process changes, which, together with the information supplied, were summarized in a posting from the group leader to the group a few days after the face-to-face meeting. One reply followed, acknowledging this posting.

The group leader left Waikato right after the group was completed; her successor responsible for coordinating the implementation of the business process changes was agreed on by the group.

Three months after the group was completed, one of the business process changes had been fully implemented, and the other two had been partially implemented.

Results

An unstructured interview with the new international students' support department manager, conducted approximately three months after the group completion, indicated satisfaction with the outcome of the business process change implementation. In her view, it had led to a moderate improvement in the quality of the services provided by her department. Unstructured interviews with a few international graduate students confirmed this perception.

Waikato.G6: New International Students

Motivation

Approximately eight months before this business process improvement group began, a system whereby senior students helped newly arrived international students with their cultural and educational adaptation was put in place by the international students office, mentioned in Waikato.G5, which had now been renamed "International Office." This system was called the "buddy system" and was based on the idea that local senior students could pass on their experience in dealing with life at Waikato and social life matters in general to international students and, at the same time, get valuable input from other cultures.

The buddy system involved both New Zealand and senior international students who helped newly arrived international students. In the buddy system, which was seen as a type of club, the local mentors were called "local buddies," and the newly arrived international students were called "international buddies."

There were some problems with this system, which were seen as needing urgent attention since the buddy system was perceived by some of Waikato's top managers as having the potential to give the institution an edge over other universities in the Australasian region.

One of the problems was that there were not enough local buddies; only six volunteers were listed by the time this business process improvement group started. In addition, participation of native New Zealand students in the buddy system was low. A number of the local students assigned to the buddy system had the same cultural background and, in some cases, spoke the same native language as the new international students. Although this was seen as likely to reduce the initial cultural shock experienced by international buddies, it was often seen by international buddies themselves as an obstacle for them to have a deeper contact with a truly different culture and to improve their English skills. Finally, there was a perceived lack of initiative from local buddies to identify and meet the needs of the international buddies who, in turn, often did not know how to approach and get help from their local buddies.

Formation

This group was a result of a request made to me by one of the members of the International Office mentioned in Waikato.G5 who had participated in that group as a member. This prospective leader had recently been assigned a middle management position in his office, and one of his responsibilities was, in his words, "to make the buddy system work ...."

This former member of a previous group wanted to lead, with my facilitation, a business process improvement group in the mid-semester break to improve the business process involved in providing support to new international students through the buddy system. He decided to invite the director of his area, his manager, local buddies, and international buddies to participate in the business process improvement group discussion. By doing so, he expected to have a wide range of opinions represented in the group discussion. He also wanted to bring the voice of the customers (i.e., the international buddies) of the business process into the discussion.

Features

The group lasted 54 days and involved 11 members from four departments (student departments were considered those where the students were doing their majors). From the 11 members who agreed to participate, only seven contributed postings to the discussion. The interaction in the group comprised 15 postings, a number of one-on-one e-mail messages, some one-on-one oral conversations, and one group face-to-face meeting. The group face-to-face meeting was conducted as part of the analysis stage. It lasted approximately two hours and involved only five members who had previously contributed postings.

According to estimates provided by the group members, 75% of the total time spent by them in the group discussion was in interactions through the e-collaboration system; the other 35% was spent in oral interactions.

Stages

In the definition stage, the leader posted a message to the group containing a list with six problems and four business processes, presented as the possible sources of the problems. These business processes were related to local buddy selection criteria, guidelines to be followed by these students when dealing with international buddies, and the communication between international office staff, local buddies, and international buddies.

The leader's initial posting was followed by 11 replies showing strong agreement about the lack of enough effective communication channels between local and international buddies. A number of these postings already included suggestions on how to improve this situation. A posting from the group leader summarized these contributions and started the analysis stage. In this posting, the group leader described the general business process of providing support to international buddies by discussing several characteristics of the buddy system, rather than following either the workflow or the data flow formats suggested in MetaProi's guide. He proposed three main changes in connection with this general business process. His posting was followed by only one reply praising the initiative and emphasizing the need for more efficient and effective communication channels between local and international buddies.

The lack of more participation prompted the leader to contact business process improvement group members, both over the phone and face-to-face, and invite them to take part in a face-to-face group meeting. This meeting was eventually conducted with only five of the business process improvement group members who agreed to participate. It lasted, as mentioned before, approximately two hours. In this meeting, business process improvement group members agreed on four business process changes, which were summarized in a posting from the group leader to the group a few days after. No replies followed.

Three months after the group completion, all business process changes were in the process of being implemented. Two of those changes had been almost fully implemented by then.

Results

One unstructured interview and a number of informal conversations with the new international students' support department manager, conducted approximately three months after the group completion, indicated satisfaction with the outcome of the business process change implementation. In her view, it led to a moderate improvement in the quality of the general business process targeted. Unstructured interviews with a few international buddies confirmed this perception.

Appendix: Statistics for Those Who Hate Statistics

The Importance of Statistical Tests

When we analyze quantitative evidence (e.g., numbers) that describe a particular situation or phenomenon, we often need to generate coefficients based on specific statistical tests to reach reasonable conclusions. Visually inspecting a table full of numbers, for example, can be quite confusing, and the related conclusions may be deceiving. This is one of the reasons why statistical tests are important. The more quantitative evidence we have to analyze, the more difficult it is to inspect it visually, and so the more important those statistical tests become.

For example, we may want to know whether a particular variable, such as the degree of e-collaboration technology use by a business process improvement group, has any effect on the duration (or lifetime) of the group, measured in days. One way of testing that is to analyze the duration of several groups, some of them conducted using e-collaboration technology support, and others conducted without any e-collaboration technology support.

By simply comparing group duration averages (also known as "means" in statistics lingo) for each condition (i.e., with and without technology support), we may find that e-collaboration technology-supported groups have, on average, a duration in days that is, for instance, 13% higher than the groups conducted without any e-collaboration technology support.

In the situation above, the following question arises. Is the 13% difference large enough to be significant? If the answer is yes, and other circumstances (e.g., group size, cultural background of the participants) were the same regarding the two group conditions (i.e., with and without technology support), then we can conclude that the use of e-collaboration technology had a significant impact on the duration of the business process improvement groups. The answer to this type of question, which is quite important in behavioral research in general, is one of the most important outcomes of statistical tests.

Statistical tests are widely used in areas other than behavioral research on the impact of technologies on people. For example, similar types of questions are whether a particular medical drug has any significant effect on individuals suffering from a certain disease, and whether a difference in the number of votes for two competing candidates in a pre-election poll is significant enough to warrant optimism in the camp of the candidate with the higher number of votes.

Three main types of statistical tests of significance used in previous chapters of this book are comparisons of means, correlation, and distribution trend tests. Comparison of means tests are aimed at establishing whether the differences between the means, or averages, of two or more conditions differ significantly from each other (as illustrated through the previous example). Correlation tests aim to establish whether two variables (e.g., degree of e-collaboration technology use and likelihood of success of a business process improvement group) vary in a significant way. Distribution trend tests aim to establish whether an observed distribution trend (e.g., the distribution of user perceptions about an e-collaboration tool's impact on group outcome quality) is significant enough to allow for the conclusion that it is caused by a particular variable (e.g., e-collaboration tool support). Each of these tests is discussed in more detail below.

Comparing Means from Different Conditions

Let us assume that we facilitated 20 business process improvement groups. Half of those groups (10 groups) used an e-collaboration system to communicate, whereas the other half communicated face-to-face. Let us also assume that the outcomes of those business process improvement groups (i.e., the

Table A.1. Outcome quality scores for 20 business process improvement groups

	Group outcome quality (scores from 1 to 7)	
Groups	Face-to-face	E-collaboration
1 and 2	4	7
3 and 4	5	6
5 and 6	3	7
7 and 8	2	5
9 and 10	5	6
11 and 12	1	2
13 and 14	4	5
15 and 16	5	7
17 and 18	3	5
19 and 20	2	6

business process redesigns generated by them) were scored in terms of quality. The scores ranged from one (very poor quality) to seven (very high quality).

Table A.1 shows the scores obtained for each of the business process improvement groups. A simple visual inspection of Table A.1 suggests that the e-collaboration technology-supported groups seem to have generally higher scores than the face-to-face groups, but a simple visual inspection usually is not enough for us to establish with certainty how much better the e-collaboration technology-supported groups did on average, and whether that difference is statistically significant.

To find out how much better the e-collaboration technology-supported groups did on average, we can calculate the mean (or average) scores obtained for both face-to-face and e-collaboration technology-supported groups. Those means are 3.4 and 5.6, respectively, which suggest that the scores for e-collaboration technology-supported groups were, on average about 65% percent higher than the scores obtained for face-to-face groups.

When we calculate means we also usually calculate standard deviations (noted as "SD" in Table A.2), which, in the example in question, are a measure of how much variation there is in the scores for each condition (i.e., face-to-face and e-collaboration). The standard deviations obtained for face-to-face and e-collaboration technology-supported groups are 1.43 and 1.51, respectively, which suggests that the degree of variability between the two conditions is similar (this is usually considered a good thing, from a statistical analysis perspective). The standard deviations can also tell us much more, but a

Table A.2. Descriptive and T test statistics

	Face-to-Face	E-Collaboration
Mean	3.40	5.60
SD	1.43	1.51

T coefficient	*3.349*
P	*0.004*

discussion about that would be somewhat technical and beyond the scope of this appendix (whose goal is not to induce readers to hate statistics even more than they already may).

Table A.2 allows us to establish whether the 65% difference between the mean scores obtained for face-to-face and e-collaboration technology-supported groups is significant from a statistical standpoint. For that, we can use a variety of comparison of means tests, of which one of the most common is the T test. A typical T test will generate two coefficients—a T coefficient and a P coefficient—which are 3.349 and .004, respectively, in the example discussed here.

Even though the test is called a T test, it is the P coefficient that really matters most, because that coefficient is the probability that the 65% difference between the mean scores obtained for face-to-face and e-collaboration technology-supported groups is due to chance. In our example, the P coefficient is .004, which means that the probability that the difference between the mean scores is due to chance is .4% (less than half of 1%). In most statistical tests, a chance probability below 5% is considered very low, and an indication that the effect under consideration is statistically significant. Therefore, in our example, the .4% probability allows us to safely say that the 65% difference between the mean scores obtained for face-to-face and e-collaboration technology-supported groups is *not* due to chance. In other words, we can safely say that the use of the e-collaboration system had a significant and positive effect on the quality of the outcomes generated by the business process improvement groups.

I know that this may sound like a convoluted and complicated way of stating the obvious. Incidentally, that is why many people hate statistics. But it is important to stress that the method behind the procedure just discussed has been developed very carefully, and is used widely in a variety of areas. One good example is the pharmaceutical industry. To prove that a vaccine is effective against a certain disease, the developer of the vaccine has to test it in a group of individuals who usually are paid to voluntarily participate in experiments involving the administration of the vaccine (other ethical considerations may exist, such as whether health risks are involved in either getting or not getting the vaccine, under the test circumstances).

One particularly convincing type of test might involve a group of individuals taking the vaccine, while another group (of about the same number) might take a placebo (i.e., a drug containing no active ingredients). If a comparison of means test (i.e., the T test) indicates that there is a statistically significant difference in the average resistance to the disease in favor of those who were administered the vaccine (when compared with those who were administered the placebo), than the pharmaceutical company that developed the vaccine strikes gold.

There are many types of comparison of means tests, and there are many statistical software packages that allow one to run those tests. Examples of other fairly widely used comparison of means tests are the one-way ANOVA and Mann-Whitney U tests. The widely used T test, which was illustrated previously, can be run on many commercial spreadsheet packages, including Microsoft Excel, which was used to generate the statistical coefficients above.

Checking for Correlations Between Variables

There is another way to test the statistical significance of the impact of the e-collaboration technology support on the quality of the outcomes generated by business process improvement groups, which was tested in the previous section through a T test. Namely, we can calculate the correlation between two variables—the degree to which e-collaboration technology support was available, and the group outcome quality scores. The former variable, the degree to which e-collaboration technology support was available, can have basically two values—1 for no support (face-to-face groups), and 2 for some support (e-collaboration technology-supported groups). The group outcome quality scores are the same as in the previous section.

Table A.3 shows the scores for the two variables that are tested for correlation using one of the most common correlation tests—the Pearson correlation test.

Table A.3. Testing an effect through a Pearson correlation test

Group	E-Collaboration Support	Outcome Quality
1	1	4
2	1	5
3	1	3
4	1	2
5	1	5
6	1	1
7	1	4
8	1	5
9	1	3
10	1	2
11	2	7
12	2	6
13	2	7
14	2	5
15	2	6
16	2	2
17	2	5
18	2	7
19	2	5
20	2	6

r coefficient	*0.620*
P coefficient	*0.004*

The results of the Pearson correlation test are shown at the bottom of Table A.3. They are the *r* coefficient, and the related *P* coefficient.

For most statistical analysis purposes, an *r* coefficient that is generated through a Pearson correlation test, and that is higher than .6, is generally seen as an indication of a strong correlation between two variables. This is consistent with the low *P* coefficient of .004 obtained (the same as in the T test employed in the previous section), which suggests that the strong correlation suggested by the Pearson correlation test has a .4% probability of being due to chance (which, again, is substantially lower than the 5% threshold used to draw conclusions from most statistical tests).

In summary, the degree to which e-collaboration technology support was available seems to have strongly and positively affected business process improvement group outcome quality in this example. If the Pearson correlation coefficient had been negative, then we could conclude the opposite; that is, that the degree to which e-collaboration technology support was available strongly and *negatively* affected group outcome quality.

Generally speaking, two variables are highly correlated when an *x-y* graph (i.e., a bi-dimensional graph) plotting their values looks like a line; in such a graph, the *x* values would be those of one of the variables, and the *y* values would be those of the other variable. The more similar to a line the graph is, the higher is the correlation between the variables. Conversely, the less similar to a line the graph is (i.e., the more dispersed the *x-y* intersection points are), the lower is the correlation between the variables.

Figure A.1. Examples of high and low correlations between variables

Figure A.1.a. High correlation (Pearson r = .98)

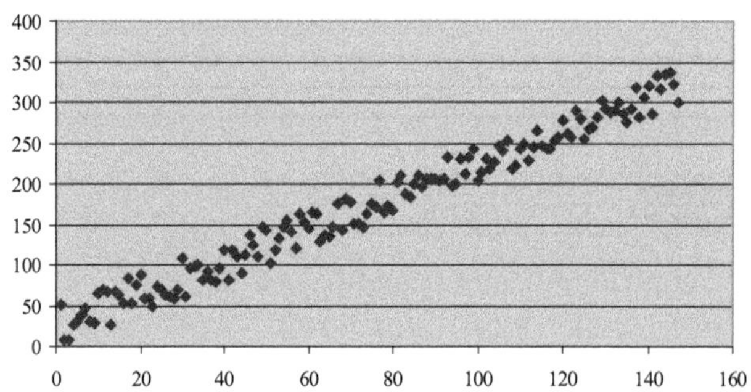

Figure A.1.b. Low correlation (Pearson r = .08)

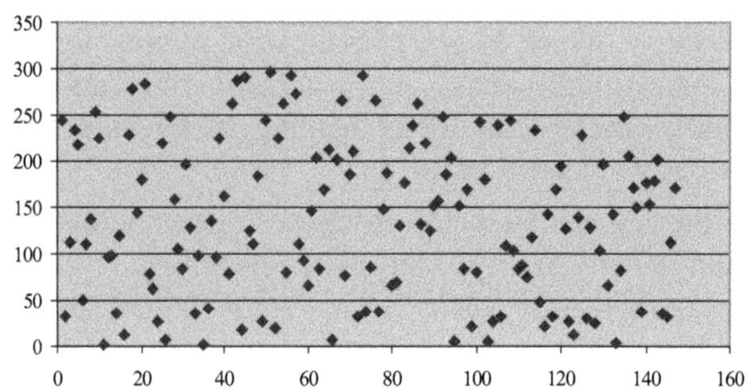

Figure A.1 illustrates this relationship. The graph at the top (Figure A.1.a) plots the relationship of two highly correlated variables whose Pearson correlation coefficient is .98. The maximum Pearson correlation coefficient possible is 1, which would be obtained if the relationship between two variables was completely linear, and which would, in turn, make the graph look like a perfect line. The graph at the bottom (Figure A.1.b) plots the relationship of two variables whose correlation is low, with a Pearson correlation coefficient of only .08.

There are several different types of correlation tests, although not as many as there are comparison of means tests, and, as with other statistical tests, there are many statistical software packages that allow one to run those different correlation tests. The widely used Pearson correlation test, which was illustrated above, can also be run on many commercial spreadsheet packages, including MS Excel—which was used to generate the statistical coefficients and the illustrative graphs above.

Assessing the Significance of Distribution Trends

Let us now consider a different type of research question that was asked and answered several times in previous chapters of this book. The question is this: How can we establish whether a trend in a distribution of perceptions regarding a single variable is due to chance? In this case, we do not have two different variables to correlate or two different conditions to compare, which prevents us from using comparison of means or correlation tests.

For example, let us assume that 100 individuals routinely conduct business process improvement group discussions by interacting in smaller groups through face-to-face meetings. Those 100 individuals then participate in e-collaboration technology-supported business process improvement group discussions in groups of similar size. Following that, they are asked whether the e-collaboration technology support decreased, had no effect, or increased the quality of the outcomes generated by their business process improvement groups.

The answers provided by the 100 individuals are distributed according to Figure A.2. As it can be seen, there seems to be an underlying trend toward

Figure A.2. Distribution of answers suggesting a strong trend (Chi-squared = 13.9, P = .0009)

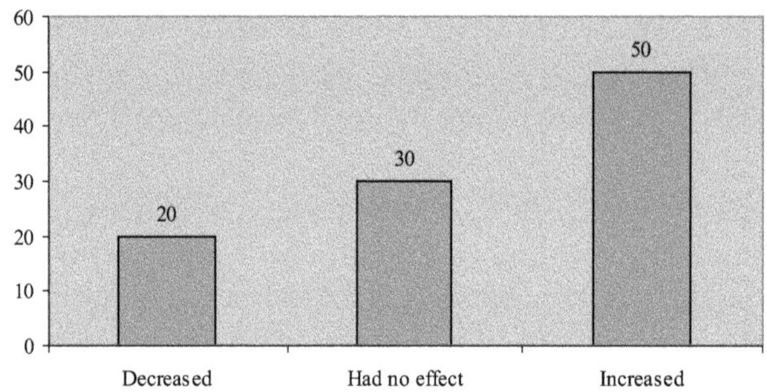

perceiving e-collaboration technology support as having increased business process improvement group outcome quality. More individuals (namely 50) perceived group outcome quality as having been increased by e-collaboration technology support than those who perceived outcome quality as not having been affected (30 individuals), or having been decreased (20 individuals). One of the ways to test the significance of that trend is to run a Chi-squared test comparing the observed distribution with a distribution in which there was no clear trend; that is, a distribution in which the same number of individuals perceived e-collaboration technology support as having increased, had no effect, or decreased group outcome quality. That number is 100 divided by 3, or approximately 33 individuals. The results of the Chi-squared test are shown near the caption of Figure A.2 (Chi-squared = 13.9, P = .0009), and suggest that the probability that the observed distribution trend is due to chance is only .09% (much lower than the 5% threshold used to draw conclusions from most statistical tests).

The conclusion that can be drawn based on this discussion and on the distribution trend suggested by Figure A.2, is that e-collaboration technology support strongly and positively affected business process improvement group outcome quality.

Figure A.3. Distribution of answers suggesting a weak trend (Chi-squared = 3.5, P = .17)

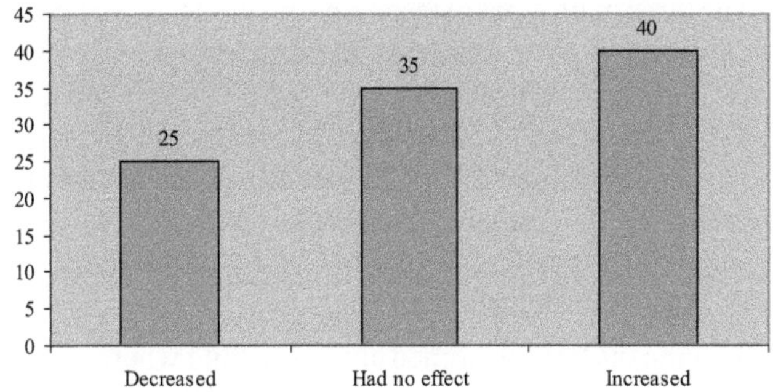

If the trend was not as skewed toward the "increased" perception, the Chi-squared test would not have been as conclusive. For example, the distribution of answers shown in Figure A.3 would still suggest a perception trend that is obviously leaning toward a general perception that group outcome quality was increased by e-collaboration technology support. However, the trend is weaker than that shown in Figure A.2, which is indicated by the Chi-squared test results near the caption of Figure A.3. Those results suggest that the probability that the observed trend is due to chance is 17%, which is too high when compared with the 5% level suggested by statisticians as the upper limit used to draw conclusions from most statistical tests. In other words, a distribution of perceptions like the one in Figure A.3 would not allow us to conclude with much certainty that e-collaboration technology support positively affected business process improvement group outcome quality.

There are a few types of tests that can be used to analyze distribution trends. As with the other statistical tests discussed in this chapter and used in previous chapters of this book, there are many statistical software packages that allow one to run those distribution trend significance tests. The relatively widely used Chi-squared test that was illustrated previously, also can be run on many commercial spreadsheet packages, including Microsoft Excel. That package was used to generate the Chi-squared and *P* coefficients above, as well as the graphs used to illustrate the distribution trends used in the Chi-squared tests.

Summary and Concluding Remarks

In previous chapters of this book, the results of several statistical tests were discussed. The three main types of tests employed were comparison of means, correlation, and distribution trend tests. A typical statistical test will generate two main coefficients—a test coefficient, which is usually named after the test (e.g., T coefficient, for a T test), and a *P* coefficient. It is usually the *P* coefficient that matters most, because it indicates the probability of chance of the test and whether the yielded result is significant or not. In most statistical tests, a *P* below 5% (i.e., below .05) is considered very low and an indication that the effect under consideration is statistically significant.

Comparison of means tests are aimed at establishing whether the difference between the means (or averages) of two or more conditions differ significantly from each other. One of the most common comparison of means tests is the T test. There are many types of comparison of means test, and there are many statistical software packages that allow one to run those tests. Examples of other fairly widely used comparison of means tests are the Mann-Whitney U and the one-way ANOVA tests.

Correlation tests aim to establish whether two variables (e.g., degree of e-collaboration technology use and likelihood of success of a business process improvement group) vary together in a significant way. One of the most common correlation tests is the Pearson correlation test, which generates an *r* coefficient and a *P* coefficient. If an *r* coefficient is higher than .6, this is generally seen as an indication of a strong correlation between the two variables under consideration. When two variables are highly correlated, an *x-y* graph plotting their values will look like a line. The more similar to a line the graph is, then the higher is the correlation between the variables; the less similar to a line the graph is, the lower is the correlation between the variables.

Distribution trend tests aim to establish whether an observed distribution trend (e.g., the distribution of user perceptions about an e-collaboration tool's impact on group outcome quality) is significant enough to allow for the conclusion that it is caused by a particular variable (e.g., e-collaboration tool support). One of the ways to test the significance of a distribution trend is to run a Chi-squared test comparing the observed distribution with a distribution in which there was no clear trend.

There are many statistical software packages that allow one to run the several statistical tests discussed in this chapter. One such statistical analysis package is SPSS, which is commercialized by a company of the same name. The T, Pearson correlation, and Chi-squared tests, which have been discussed in this chapter, can also be run to a large extent on many commercial spreadsheet packages, including Microsoft Excel.

Most statistical tests, including the ones discussed in this chapter, are best run when what is known as the "sample size" is relatively large; otherwise, they lose their power. For example, let us assume that we want to test the statistical significance of the impact of the e-collaboration technology support on the quality of the outcomes generated by business process improvement groups. It will be better to run a correlation test based on evidence collected from 50 groups than on evidence collected from only 10 groups. That is, the former will yield more reliable results than the latter. Conversely, the larger the sample size, the less strong the underlying effect needs to be to yield a statistically significant result, which means that with very large samples, even weak effects will be statistically significant.

References

Alster, N. (1997, April 3). Do US firms spend too much on information technology? *Investor's Business Daily*.

Argyris, C. (1977, September-October). Double loop learning organizations. *Harvard Business Review*,

Argyris, C. (1992). *On organizational learning*. Cambridge, MA: Blackwell.

Argyris, C., & Schon, D. (1978). *Organizational learning: A theory of action perspective*. Reading, MA: Addison-Wesley.

Asper, G., Lima, L.A., & Kock, N. (1999). Redesigning critical supply chain processes: A health care experience. *Proceedings of the 4th International Symposium on Logistics*, Padova, Italy.

Asper, G., Santana, C., & Kock, N. (2001). Strategies for building networks: Toward a Brazilian health care value chain. *Proceedings of the 7th International Symposium on Logistics*, Padova, Italy.

Bannon, L.J. (1993). CSCW: An initial exploration. *Scandinavian Journal of Information Systems*, *5*(1).

Bartlett, F.C. (1932). *Remembering*. Cambridge, MA: Cambridge University Press.

Bartlett, M. (2001). Good e-mail communication requires hard work. [Electronic version]. (*washingtonpost.com*). Retrieved from *http://www.newsbytes.com/news/01/168549.html*

Bashein, B.J., & Markus, M.L. (1994). Preconditions for BPR success. *Information Systems Management, 11*(2).

Bates, B., & Cleese, J. (2001). *The human face*. New York: DK Publishing.

Biggs, M. (2000). Enabling a successful e-business strategy requires a detailed business process map. *InfoWorld, 22*(10).

Boaz, N.T., & Almquist, A.J. (1997). *Biological anthropology: A synthetic approach to human evolution.* Upper Saddle River, NJ: Prentice Hall.

Boland Jr., R.J., & Tenkasi, R.V. (1995). Perspective making and perspective taking in communities of knowing. *Organization Science, 6*(4).

Booch, G., Jacobson, I., & Rumbaugh, J. (1998). *The unified modeling language user guide.* New York: Addison-Wesley.

Boroson, W. (1997). *Keys to investing in mutual funds.* Hauppauge, NY: Barron's Educational Series.

Brynjolfsson, E., & Hitt, L.M. (1998). Beyond the productivity paradox. *Communications of the ACM, 41*(8).

Buchanan, J., & Kock, N. (2000). Information overload: A decision making perspective. *Proceedings of the 15th International Conference on Multiple Criteria Decision Making*, Berlin, Germany.

Budd, J.M., & Raber, D. (1996). Discourse analysis: Method and application in the study of information. *Information Processing & Management, 32*(2).

Burke, G., & Peppard, J. (Eds) (1995). *Examining business process re-engineering.* London: Kogan Page.

Burke, K., & Chidambaram, L. (1999). How much bandwidth is enough? A longitudinal examination, of media characteristics and group outcomes. *MIS Quarterly, 23*(4).

Burn, J., & Barnett, M. (1999). Communicating for advantage in the virtual organization. *IEEE Transactions on Professional Communication, 42*(4).

Buzacott, J.A. (1996). Commonalities in reengineered business processes: Models and issues. *Management Science, 42*(5).

Caldwell, B. (1994, June 20). Missteps, miscues. *InformationWEEK.*

Callatay, A.M. (1986). *Natural and artificial intelligence.* North-Holland, Amsterdam: The Netherlands.

Camerer, C.F., Johnson, E.J. (1991). The process-performance paradox in expert judgment. In K.A. Ericsson, & J. Smith (Eds.), *Toward a General Theory of Expertise*. Cambridge, MA: Cambridge University Press.

Carlson, J.R., & Zmud, R.W. (1999). Channel expansion theory and the experiential nature of media richness perceptions. *Academy of Management Journal*, *42*(2).

Caron, J.R., Jarvenpaa, S.L., & Stoddard, D.B. (1994). Business reengineering at CIGNA Corporation: Experiences and lessons learned from the first five years. *MIS Quarterly*, *18*(3).

Carr, N.G. (2003). IT doesn't matter. *Harvard Business Review*, *81*(5).

Carr, N.G. (2003a). IT? Does it matter? *Network Magazine*, *18*(7).

Cartwright, J. (2000). *Evolution and human behavior: Darwinian perspectives on human nature*. Cambridge, MA: MIT Press.

Champy, J. (1995). *Reengineering management*. New York: Harper Business.

CHE (1995, March). The learning organization. *Chief Executive*.

Checkland, P. (1981). *Systems thinking, systems practice*. New York: John Wiley & Sons.

Checkland, P., & Scholes, J. (1990). *Soft systems methodology in action*. New York: John Wiley & Sons.

Childe, S.J. (1995). Business process re-engineering. *Management Bibliographies & Reviews*, *21*(3).

Childe, S.J., Maull, R.S., & Benett, J. (1994). Frameworks for understanding business process re-engineering. *International Journal of Operations & Productions Management, 14*(12).

Choi, T.Y., & Behling, O.C. (1997). Top managers and TQM success: One more look after all these years. *The Academy of Management Executive, 11*(1).

Choi, T.Y., & Liker, J.K. (1995). Bringing Japanese continuous improvement approaches to U.S. manufacturing: The roles of process orientation and communications. *Decision Sciences*, *26*(5).

Chuang, T., & Yadav, S.B. (2000). A decision-driven approach to object-oriented analysis. *Database for Advances in Information Systems*, *31*(2).

Clemons, E.K., Thatcher, M.E., & Row, M.C. (1995). Identifying sources of reengineering failures: A study of the behavioral factors contributing to reengineering risks. *Journal of Management Information Systems, 12*(2).

Crosby, P.B. (1980). *Quality is free: The art of making quality certain.* New York: Mentor.

Crosby, P.B. (1984). *Quality without tears: The art of hassle-free management.* New York: McGraw-Hill.

D'Arcy, J., & Kock, N. (2003). Unraveling the e-collaboration paradox: Evidence of compensatory adaptation to low media naturalness. *Proceedings of the 13th Information Resources Management International Conference.* Hershey, PA: Idea Group Publishing.

Daft, R.L., & Lengel, R.H. (1986). Organizational information requirements, media richness and structural design. *Management Science, 32*(5).

Daft, R.L., Lengel, R.H., & Trevino, L.K. (1987). Message equivocality, media selection, and manager performance: Implications for information systems. *MIS Quarterly, 11*(3).

Danesh, A., Kock, N., & Liu, C. (2003). Effectiveness of utilizing information flow modeling in an information-based organization. *Proceedings of the 2003 International Applied Business Research Conference*, Littleton, CO.

Davenport, T.H. (1993). *Process innovation.* Boston: Harvard Business Press.

Davenport, T.H. (1993a). Need radical innovation and continuous improvement? Integrate process re-engineering and total quality management. *Planning Review, 21*(3).

Davenport, T.H. (2000) *Mission critical: Realizing the promise of enterprise systems*. Boston: Harvard Business School Press.

Davenport, T.H., & Beck, J.C. (2001). *The attention economy: Understanding the new currency of business.* Cambridge, MA: Harvard Business School Publishing.

Davenport, T.H., & Prusak, L. (2000). *Working knowledge.* Cambridge, MA: Harvard Business School Press.

Davenport, T.H., & Short, J.E. (1990, Summer). The new industrial engineering: Information technology and business process redesign. *Sloan Management Review.*

Davenport, T.H., & Stoddard, D.B. (1994). Reengineering: Business change of mythic proportions? *MIS Quarterly*, *18*(2).

Davenport, T.H., Jarvenpaa, S.L., & Beers, M.C. (1996). Improving knowledge work processes. *Sloan Management Review*, *37*(4).

Davidow, W.H., & Malone, M.S. (1992). *The virtual corporation*. New York: HarperCollins.

Davis, W.S. (1983). *System analysis and design: A structured approach*. Reading, MA: Addison-Wesley.

Dean, D.L., Lee, J.D., Orwig, R.E., & Vogel, D.R. (1995). Technological support for group process modeling. *Journal of Management Information Systems, 11*(3).

DeLuca, D., & Kock, N. (2000). Improving business processes asynchronously. *Proceedings of the 10th Information Resources Management International Conference*. Hershey, PA: Idea Group Publishing.

DeLuca, D.C. (2003). *Business process improvement using asynchronous e-collaboration: Testing the compensatory adaptation model*. [Ph.D. Thesis]. Philadelphia, PA: Temple University.

DeMarco, T. (1979). *Structured analysis and system specification*. Englewood Clilffs, NJ: Prentice-Hall.

Deming, W.E. (1986). *Out of the crisis*. Cambridge, MA: MIT, Center for Advanced Engineering Study.

Dennett, D.C. (1991). *Consciousness explained*. Boston: Little, Brown and Co.

Dennis, A., & Wixom, B.H. (2000). *Systems analysis and design: An applied approach*. New York: John Wiley & Sons.

Dennis, A.R., Haley, B.J., & Vanderberg, R.J. (1996). A meta-analysis of effectiveness, efficiency, and participant satisfaction in group support systems research. *Proceedings of the 17th International Conference on Information Systems*, New York.

Dennis, A.R., Hayes, G.S., & Daniels, R.M., Jr. (1999). Business process modeling with group support systems. *Journal of Management Information Systems*, *15*(4).

Dewan, S., & Kraemer, K.L. (1998). International dimensions of the productivity paradox. *Communications of the ACM*, *41*(8).

Dingle, M.E. (1994) *Business process reengineering: A New Zealand perspective*. [Research Report]. Palmerston North, New Zealand: Department of Executive Education, Massey University.

Dobzhansky, T. (1971). *Mankind evolving: The evolution of the human species*. New Haven, CT: Yale University Press.

Dos Santos, B.L., Peffers, K., & Mauer, D.C. (1993). The impact of information technology investment announcements on the market value of the firm. *Information Systems Research*, *4*(1).

Dozier, R.W., Jr. (1992). *Codes of evolution*. New York: Crown Publishers.

Drucker, P.F. (1989). *The new realities* New York: Harper & Row.

Drucker, P.F. (1995, February). Rethinking work. *Executive Excellence*.

Earl, M.J. (1994). The new and the old of business process redesign. *Journal of Strategic Information Systems*, *3*(1).

Eason, K. (1996). Division of labour and the design of systems for computer support for cooperative work. *Journal of Information Technology*, *11*(1).

El-Shinnawy, M., & Markus, L. (1998). Acceptance of communication media in organizations: Richness or features? *IEEE Transactions on Professional Communication*, *41*(4).

Feldman, J. (1986) On the difficulty of learning from experience. In H.P. Sims, Jr., & D.A. Gioia (Eds.), *The Thinking Organization*. San Francisco: Jossey-Bass.

Fingar, P., Aronica, R., & Maizlish, B. (2001). *The death of "e" and the birth of the real new economy*. Tampa, FL: Meghan-Kiffer Press.

Galbraith, J.K. (1973). *Designing complex organizations*. Reading, MA: Addison-Wesley.

Gane, C., & Sarson, T. (1979). *Structured systems analysis: Tools and techniques*. Englewood Cliffs, NJ: Prentice-Hall.

Gardner, H. (1985). *The mind's new science*. New York: Basic Books.

Garratt, R. (1994). *The learning organization*. London: HarperCollins.

Geyelin, M. (1994, August 8). Doomsday device. *The Wall Street Journal*.

Glassman, R.B. (2003). Topology and graph theory applied to cortical anatomy may help explain working memory capacity for three or four simultaneous items. *Brain Research Bulletin*, *60*(1).

Gleick, J. (1993). *Chaos: Making a new science*. London: Abacus.

Goldratt, E.M. (1991). *The haystack syndrome: Sifting information out of the data ocean*. New York: North River Press.

Goldratt, E.M., & Fox, R.E. (1986). *The race*. New York: North River Press.

Gore, M., & Stubbe, J.W. (1988). *Elements of systems analysis*. Dubuque, IA: Brown Publishers.

Graves, R.H. (2001). Seeking defense efficiency. *Acquisition Review Quarterly, 8*(3).

Green, S.B., Salkind, N.J., & Akey, T.M. (1997). *Using SPSS for windows: Analyzing and understanding data*. Upper Saddle River, NJ: Prentice Hall.

Grover, V., Jeong, S.R., Kettinger, W.J., & Teng, J.T.C. (1995). The implementation of business process reengineering. *Journal of Management Information Systems, 12*(1).

Grudin, J. (1994). Groupware and social dynamics: Eight challenges for developers. *Communications of the ACM, 37*(1).

Grudin, J. (1994a). CSCW: History and focus. *IEEE Computer, 27*(5).

Guha, S., Kettinger, W.J., & Teng, J.T.C. (1993, Summer). Business process reengineering, building a comprehensive methodology. *Information Systems Management*.

Hackett, P. (1990, Winter). Investment in technology – The service sector sinkhole? *Sloan Management Review*.

Hall, G., Rosenthal, J., & Wade, J. (1993, November-December). How to make reengineering really work. *Harvard Business Review*.

Hammer, M. (1990, July-August). Reengineering work: Don't automate, obliterate. *Harvard Business Review*.

Hammer, M.E. (1996). *Beyond reengineering*. New York: HarperCollins.

Hammer, M. (2000). Reengineering redux. *CIO Magazine, 13*(10).

Hammer, M., & Champy, J. (1993). *Reengineering the corporation*. New York: Harper Business.

Hammer, M., & Stanton, S.A. (1995). *The reengineering revolution*. New York: HarperCollins.

Hammer, M., & Stanton, S.A. (1997). The reengineering revolution. *Government Executive, 27*(9).

Harrington, H.J. (1991). *Business process improvement*. New York: McGraw-Hill.

Harrington, H.J., Esseling, E.K.C., & Van Nimwegen, H. (1998). *Business process improvement workbook: Documentation, analysis, design, and management of business process improvement*. New York: McGraw-Hill.

Hawking, S.W. (1988). *A brief history of time*. New York: Bantam Books.

Hayek, F.A. (1996). The use of knowledge in society. In P.S. Myers (Ed.), *Knowledge management and organizational design*. Boston: Butterworth-Heinemann.

Hendricks, K.B., & Singhal, V.R. (1997). Does implementing an effective TQM program actually improve operating performance? Empirical evidence from firms that have won quality awards. *Management Science*, *43*(9).

Herzberg F., Mausner, B., & Snyderman, B.B. (1959). *The motivation to work*. New York: Wiley.

Hiltz, S.R. (1978). Controlled experiments with computerized conferencing: Results of a pilot study. *Bulletin of the American Society for Information Science*, *4*(5).

Hirschheim, R.A. (1985). *Office automation: A social and organizational perspective*. New York: John Wiley & Sons.

Holsapple, C.W., & Whinston, A.B. (1996). *Decision support systems: A knowledge-based approach*. St. Paul, MN: West Publishing.

Holyoak, K.J. (1991). Symbolic connectionism: Toward third-generation theories of expertise. In K.A. Ericsson, & J. Smith (Eds.), *Toward a general theory of expertise*. Cambridge, MA: Cambridge University Press.

Hunt, V.D. (1996). *Process mapping: How to reengineer your business processes*. New York: John Wiley & Sons.

Hurtado, P.S. (1998). A comparative analysis of traditional cognitivistic and autopoietic systems theory perspectives of orthodox problem-solving/decision-making approaches. *Proceedings of the Decision Sciences Institute Conference*, Atlanta, GA.

Huttunen, K., Hoover, B., & Eloranta, E. (2001). *Managing the demand-supply chain*. New York: John Wiley & Sons.

Jacobson, I., Ericsson, M., & Jacobson, A. (1995). *The object advantage.* New York: Addison-Wesley.

Jessup, L.M., & Tansik, D.A. (1991). Decision making in an automated environment: The effects of anonymity and proximity with a group decision support system. *Decision Sciences*, *22*(2).

Jung, C.G. (1968). *Analytical psychology: Its theory and practice*. New York: Vintage Books.

Jung, C.G. (1989). *Memories, dreams, reflections*. New York: Vintage Books.

Juran, J. (1989). *Juran on leadership for quality*. New York: The Free Press.

Katzenbach, J.R., & Smith, D.K. (1993). *The wisdom of teams: Creating the high-performance organization*. Boston: Harvard Business School Press.

Katzenstein, G., & Lerch, F.J. (2000). Beneath the surface of organizational processes: A social representation framework for business process redesign. *ACM Transactions on Information Systems*, *18*(4).

Kettinger, W.J., & Grover, V. (1995). Toward a theory of business change management. *Journal of Management Information Systems*, *12*(1).

Kock, N. (1995). *MetaProi: A broup process for business process improvement*. [Project Report GP-G-1995-R5]. Hamilton, New Zealand: University of Waikato.

Kock, N. (1995a). *Process reengineering, PROI: A practical methodology*. Sao Paulo, Brazil: Editora Vozes.

Kock, N. (1997). Fostering interdepartmental knowledge communication through groupware: A process improvement perspective. *Proceedings of the International Conference on Supporting Group Work*, New York.

Kock, N. (1998). Can communication medium limitations foster better group outcomes? An action research study. *Information & Management*, *34*(5).

Kock, N. (1998a). Government transformation and structural rigidity: Redesigning a service acquisition process. *Acquisition Review Quarterly*, *5*(1).

Kock, N. (1999). *Process improvement and organizational learning: The role of collaboration technologies*. Hershey, PA: Idea Group Publishing.

Kock, N. (2000). Information overload: A process-centered view. *Knowledge and Process Management*, *7*(4).

Kock, N. (2000a). Benefits for virtual organizations from distributed groups. *Communications of the ACM*, *43*(11).

Kock, N. (2001). Asynchronous and distributed process improvement: The role of collaborative technologies. *Information Systems Journal*, *11*(2).

Kock, N. (2001a). Changing the focus of business process redesign from activity flows to information flows: A defense acquisition application. *Acquisition Review Quarterly*, *8*(2).

Kock, N. (2001b). Compensatory adaptation to a lean medium: An action research investigation of electronic communication in process improvement groups. *IEEE Transactions on Professional Communication*, *44*(4).

Kock, N. (2001c). The ape that used email: Understanding e-communication behavior through evolution theory. *Communications of the AIS*, *5*(3).

Kock, N. (2001d). From apes to customers: In search of the e-communication gene. *Proceedings of the IFIP WG 8.2 Workshop on Organizations and Society in Information Systems*, Syracuse, NY.

Kock, N. (2002). Managing with Web-based IT in mind. *Communications of the ACM*, *45*(5).

Kock, N. (2002a). Evolution and media naturalness: A look at e-communication through a Darwinian theoretical lens. *Proceedings of the 23rd International Conference on Information Systems*, Atlanta, GA.

Kock, N. (2002b). Human evolution, genes and e-communication in organizations. *Proceedings of the 12th Information Resources Management International Conference*, Hershey, PA.

Kock, N. (2002c). *Compensatory adaptation: Understanding how obstacles can lead to success.* Haverford, PA: Infinity Publishing.

Kock, N. (2003). Communication-focused business process redesign: Assessing a communication flow optimization model through an action research study at a defense contractor. *IEEE Transactions on Professional Communication*, *46*(1).

Kock, N., & Corner, J.L. (1997). Improving university processes through computer-mediated process redesign groups. *Campus-Wide Information Systems*, *14*(1).

Kock, N., & D'Arcy, J. (2002). Resolving the e-collaboration paradox: The competing influences of media naturalness and compensatory adaptation. *Information Management and Consulting (Special Issue on Electronic Collaboration), 17*(4).

Kock, N., & Danesh, A. (2003). Shifting the focus of process redesign from activity flows to communication flows: Evidence from an experimental study. *Proceedings of the 1st Conference on Systems Integration*, Hoboken, NJ.

Kock, N., & Davison, R. (2003). Can lean media support knowledge sharing? Investigating a hidden advantage of process improvement. *IEEE Transactions on Engineering Management, 50*(2).

Kock, N., & McQueen, R.J. (1995). Integrating groupware technology into a business process improvement framework. *Information Technology & People, 8*(4).

Kock, N., & McQueen, R.J. (1996). Product flow, breadth and complexity of business processes: An empirical study of fifteen business processes in three organizations. *Business Process Re-engineering & Management, 2*(2).

Kock, N., & McQueen, R.J. (1997). Using groupware in quality management programs. *Information Systems Management, 14*(2).

Kock, N., & McQueen, R.J. (1998). An action research study of effects of asynchronous groupware support on productivity and outcome quality of process redesign groups. *Journal of Organizational Computing and Electronic Commerce, 8*(2).

Kock, N., & McQueen, R.J. (1998a). Knowledge and information communication in organizations: An analysis of core, support and improvement processes. *Knowledge and Process Management, 5*(1).

Kock, N., & McQueen, R.J. (1998b). Groupware support as a moderator of interdepartmental knowledge communication in process improvement groups: An action research study. *Information Systems Journal, 8*(3).

Kock, N., & Murphy, F. (2001). *Redesigning acquisition processes: A new methodology based on the flow of knowledge and information*. Fort Belvoir, VA: Defense Acquisition University Press.

Kock, N., & Tomelin, C.A. (1996). *PMQP: Total quality management in practice*. Curitiba, Brazil: Business Development Center, SENAC.

Kock, N., Davison, R., Ocker, R., & Wazlawick, R. (2001). E-collaboration: A look at past research and future challenges. *Journal of Systems and Information Technology (Special Issue on E-Collaboration), 5*(1).

Kock, N., Hassell, L., & Wazlawick, R.S. (2002). Tacit knowledge sharing and human evolution: A framework for developing "natural" collaboration technologies. *Proceedings of the 1st ISOneWorld Conference*, Las Vegas, NV.

Kock, N., Hilmer, K., Standing, C., & Clark, S. (2000). Supporting learning processes with collaboration technologies: A brief look at past research and challenges that lie ahead. *Journal of Systems and Information Technology*, *4*(4).

Kock, N., Jenkins, A., & Wellington, R. (1999). A field study of success and failure factors in asynchronous groupware-supported process improvement groups. *Business Process Management*, *5*(3).

Kock, N., McQueen, R.J., & Baker, M. (1996). Learning and process improvement in knowledge organizations: A critical analysis of four contemporary myths. *The Learning Organization*, 3(1).

Kock, N., McQueen, R.J., & Corner, J.L. (1997). The nature of data, information and knowledge exchanges in business processes: Implications for process improvement and organizational learning. *The Learning Organization*, *4*(2).

Kofman, F., & Senge, P.M. (1993). The heart of learning organizations. *Organization Dynamics*, *22*(2).

Kogut, B., & Zander, U. (1992). Knowledge of the firm, combinative capabilities, and the replication of technology. *Organization Science*, *3*(3).

Kryt, J. (1997). Information conundrum: Semantics…with a payoff! *Informatica*, *21*(2).

Kurke, L.B., & Aldrich, H.E. (1983). Mintzberg was right! A replication and extension of the nature of managerial work. *Management Science*, *29*(8).

Kurzweil, R. (1990). *The age of intelligent machines*. Cambridge, MA: MIT Press.

Kurzweil, R. (1999). *The age of spiritual machines*. New York: Viking.

Laitman, J.T. (1984). The anatomy of human speech. *Natural History*, *20*(7).

Laitman, J.T. (1993). The anatomy of human speech. In R.L. Ciochon, & J.G. Fleagle (Eds.), *The Human Evolution Source Book*. Englewood Cliffs, NJ: Prentice-Hall.

Lee, A.S. (1994). Electronic mail as a medium for rich communication: An empirical investigation using hermeneutic interpretation. *MIS Quarterly*, *18*(2).

Lengel, R.H., & Daft, R.L. (1988). The selection of communication media as an executive skill. *Academy of Management Executive*, *2*(3).

Lewin, R. (1993, June 5). An orderly progression out of chaos. *New Scientist*.

Leyman, F., & Altenhuber, W. (1994). Managing business processes as an information resource. *IBM Systems Journal*, *33*(2).

Lieberman, P. (1998). *Eve spoke: Human language and human evolution*. New York: W.W. Norton & Company.

Lord, R.G., & Foti, R.J. (1986). Schema theories, information processing and organizational behaviour. In H.P. Sims, Jr., & D.A. Gioia (Eds.), *The thinking organization*. San Francisco: Jossey-Bass.

Malone, T.W., & Crowston, K. (1994). The interdisciplinaly study of coordination. *ACM Computing Surveys*, *26*(1).

Mark, J.A. (1984). Productivity measurement: The government's role in the United States. In P. Sherer, & T. Malone (Eds.), *The measurement and implications of productivity growth*. Canberra, Australia: Australian Government Publishing Service.

Markus, M.L. (1994). Electronic mail as the medium of managerial choice. *Organization Science*, *5*(4).

Martin, J. (1991). *Rapid application development*. New York: Macmillan.

Maturana, H.R., & Varela, F.J. (1991). *Autopoiesis and cognition: The realization of the living*. Dordrecht, Holland: Reidel Publishing.

Maull, R., Childe, S., Bennett, J., Weaver, A.M., & Smart, P.A. (1995). *Report on different types of manufacturing processes and IDEF0 models describing standard business processes* [Working Paper WP/GR/J95010-4]. Plymouth, England: School of Computing, University of Plymouth.

McNeill, D. (1998). *The face: A natural history*. Boston: Little, Brown and Company.

McQueen, R.J. (1991). *The effect of voice input on information exchange in computer supported asynchronous group communication* [Ph.D. Thesis]. Hamilton, New Zealand: University of Waikato.

McQueen, R.J., Payner, K., & Kock, N. (1999). Contribution by participants in face-to-face business meetings: Implications for collaborative technology. *Journal of Systems and Information Technology, 3*(1).

Meyer, B. (1998). The future of object technology. *IEEE Computer, 31*(1).

Miller, G.A. (1956). The magical number seven, plus or minus two: Some limits on our capacity for processing information. *The Psychological Review, 63*(1).

Mingers, J. (2002). Can social systems be autopoietic? Assessing Luhmann's social theory. *Sociological Review, 50*(2).

Mintzberg, H. (1975, July-August). The manager's job: Folklore and fact. *Harvard Business Review*.

Misterek, S.D.A., Dooley, K.J., & Anderson, J.C. (1992). Productivity as a performance measure. *International Journal of Operations & Production Management, 12*(1).

Mowshowitz, A. (1997). Virtual organization. *Communications of the ACM, 40*(9).

Nelson, K.M., & Cooprider, J.G. (1996). The contribution of shared knowledge to IS group performance. *MIS Quarterly, 20*(4).

Nevis, E.C., DiBella, A.J., & Gould, J.M. (1995, Winter). Understanding organisations as learning systems. *Sloan Management Review*.

Ngwenyama, O.K., & Lee, A.S. (1997). Communication richness in electronic mail: Critical social theory and the contextuality of meaning. *MIS Quarterly*, 21(2), 145-167.

Nissen, M.E. (1997, Winter). Reengineering the RFP process through knowledge-based systems. *Acquisitions Review Quarterly*.

Nissen, M.E. (1998). Redesigning reengineering through measurement-driven inference. *MIS Quarterly, 22*(4).

Nosek, J., & McNeese, M.D. (1997). Issues for knowledge management from experiences in supporting group knowledge elicitation and creation in ill-defined, emerging situations *Proceedings of the AIII Spring Symposium on Artificial Intelligence in Knowledge Management*, Calgary, Alberta, Canada.

Olesen, K.,, & Myers, M.D. (1999). Trying to improve communication and collaboration with information technology: An action research project which failed. *Information Technology & People*, *12*(4).

Olson, D.L., & Courtney, J.F., Jr. (1992). *Decision support models and expert systems*. New York: Macmillan.

Ong, W.J. (1988). *Orality and literacy: The technologizing of the world*. London: Routledge.

Oppenheim, P. (1992). *Japan without blinders*. Tokyo: Kodansha International.

Orlikowski, W.J. (1992). Learning from notes: Organizational issues in groupware implementation. *Proceedings of CSCW'92 Conference*, New York.

Ould, M.A. (1995). *Business processes: Modelling and analysis for re-engineering and improvement*. Chichester, UK: John Wiley & Sons.

Parente, R. (2003). *Strategic modularization in the Brazilian automotive industry: An empirical analysis of its antecedents and performance implications* [Ph.D. Thesis]. Philadelphia, PA: Temple University.

Partridge, C. (1994). Modelling the real world: Are classes abstractions or objects? *Journal of Object-Oriented Programming*, *7*(7).

Pawlowicz, D., Hantula, D.A., Kock, N.,, & D'Arcy J.P. (2003). Identifying and understanding media naturalness and temporal adaptation in virtual teams: Performance–a matter of time. *Proceedings of the Symposium on Innovations in Technology-Enhanced Employment Processes*, Syracuse, NY.

Peters, T.J., & Waterman, R.H., Jr. (1982). *In search of excellence*. New York: Harper & Row.

Pinker, S. (1994). *The language instinct*. New York: William Morrow and Co.

Pinker, S. (1997). *How the mind works*. New York: W.W. Norton & Co.

Popper, K.R. (1992). *Logic of scientific discovery*. New York: Routledge.

Porter, M.E. (1980). *Competitive strategy: Techniques for analyzing industries and competitors*. New York: The Free Press.

Porter, M.E. (1985). *Competitive advantage*. New York: The Free Press.

Prahalad, C.K., & Hamel, G.K. (1990). The core competence of the corporation. *Harvard Business Review, 68*(3).

Pressman, R. (1987). *Software engineering: A practitioner's approach.* New York: McGraw-Hill.

Probst, G.J.B., & Buchel, B.S.T. (1997). *Organizational learning: The competitive advantage of the future.* London: Prentice Hall.

Rai, A., Patnayakuni, R., & Patnayakuni, N. (1997). Technology investment and business performance. *Communications of the ACM, 40*(7).

Ramsey, D.K. (1987). *The corporate warriors.* Boston: Houghton Mifflin.

Redding, J.C., & Catalanello, R.F. (1994). *Strategic readiness: The making of the learning organization.* San Francisco: Jossey-Bass.

Rigby, D. (1993). The secret history of process reengineering. *Planning Review, 21*(2).

Rogers, Y. (1992). Ghosts in the network: Distributed troubleshooting in a shared working environment *Proceedings of CSCW'92 Conference,* New York.

Rose, M. (1998, January 28). Germany's SAP posts 72% jump in pretax profit. *Wall Street Journal.*

Rosenthal, R., & Rosnow, R.L. (1991). *Essentials of behavioral research: Methods and data analysis.* Boston: McGraw-Hill.

Roskelly, H. (1994). The risky business of group work. In G. Tate, E. Corbett, & N. Myers (Eds.), *The writing teacher's sourcebook.* New York: Oxford University Press.

Ruddell, S., & Stevens, J.A. (1998). The adoption of ISO 9000, ISO 14001, and the demand for certified wood products in the business and institutional furniture industry. *48*(3).

Rumbaugh, J., Jacobson, I., & Booch, G. (1998). *The unified modeling language reference manual.* New York: Addison-Wesley.

Russel, S., & Norvig, P. (1995). *Artificial intelligence: A modern approach.* Upper Saddle River, NJ: Prentice Hall.

Schacter, D.L. (2001). *The seven sins of memory: How the mind forgets and remembers.* New York: Houghton Mifflin.

Schmidt, K. (1994). *Modes and mechanisms of interaction in cooperative work.* Roskilde, Denmark: Risø National Laboratory.

Senge, P.M. (1990). *The fifth discipline.* New York: Doubleday.

Senge, P.M., Roberts, C., Ross, R.B., Smith, B.J., & Kleiner, A. (1994). *The fifth discipline fieldbook*. London: Nicholas Brealey.

Sheffield, J., & Gallupe, B. (1993). Using electronic meeting technology to support economic development in New Zealand: Short term results. *Journal of Management Information Systems, 10*(3).

Short, J., Williams, E., & Christie, B. (1976). *The social psychology of telecommunications*. London: John Wiley.

Simon, H.A., & Chase, W.G. (1973). Skill in chess. *American Scientist, 61*(4).

Smith, A. (1910, 1776). *The wealth of nations, Vol.1*. London: J.M. Dent & Sons.

Smith, A. (1910a, 1776). *The wealth of nations, Vol.2*. London: J.M. Dent & Sons.

Somerville, I. (1992). *Software engineering*. New York: Addison-Wesley.

Spender, J.C. (1996). Competitive advantage from tacit knowledge? Unpacking the concept and its strategic implications. In B. Moingeon, & A. Edmonson (Eds.), *Organizational learning and competitive advantage*. Thousand Oaks, CA: Sage.

Sproull, L., & Kiesler, S. (1986). Reducing social context cues: Electronic mail in organizational communication. *Management Science, 32*(1).

Sproull, L., & Kiesler, S. (1991). Computers, networks and work. *Scientific American, 265*(3).

Stacey, R.D. (1995). The science of complexity: An alternative perspective for strategic change processes. *Strategic Management Journal, 16*(1).

Stoddard, D.B., & Jarvenpaa, S.L. (1995). Business process redesign: Tactics for managing radical change. *Journal of Management Information Systems, 12*(1).

Strassmann, P.A. (1996). *The value of computers, information and knowledge* [Working Paper]. New Canaan: Strassmann Inc.

Strassmann, P.A. (1997). *The squandered computer*. New Canaan, CT: The Information Economics Press.

Strassmann, P.A. (1999). *Information productivity: Assessing information management costs of U.S. corporations*. New Canaan: The Information Economics Press.

Sucham, J., & Hayzak, G. (2001). The communication characteristics of virtual teams: A case study. *IEEE Transactions on Professional Communication, 44*(3).

Taylor, F.W. (1885). *A piece rate system*. New York: McGraw-Hill.

Taylor, F.W. (1911). *The principles of scientific management*. New York: Norton & Company.

Teichman, J., & Evans, K.C. (1995). *Philosophy: A beginner's guide*. Oxford, UK: Blackwell.

Teng, J.T.C., Seung, R.J., & Grover, V. (1998). Profiling successful reengineering projects. *Communications of the ACM, 41*(6).

Thomas, D. (1989, March). What's in an object. *Byte*.

Toffler, A. (1970). *Future shock*. New York: Bantam Books.

Toffler, A. (1991). *Powershift*. New York: Bantam Books.

Trist, E.L., Higgin, G.W., Pollock, A.E., & Murray, H.A. (1970). Sociotechnical systems, *group processes*. Middlesex, UK: Penguin Books.

Tufte, E.R. (1990). *Envisioning information*. Philadelphia, PA: Graphics Press.

Tufte, E.R. (1997). *Visual explanations: Images and quantities, evidence and narrative*. Philadelphia, PA: Graphics Press.

Turoff, M. (1973). Human communications via data networks. *Computer Decisions, 5*(1).

Turoff, M. (1975). The future of computer conferencing. *The Futurist, 9*(4).

Turoff, M. (1978). The EIES experience: Electronic information exchange system. *Bulletin of the American Society for Information Science, 4*(5).

Umar, A. (1997). *Object-oriented client/server Internet environments* Upper Saddle River, NJ: Prentice Hall.

Venkatraman, N. (1994). IT-enabled business transformation: From automation to business scope redefinition. *Sloan Management Review, 35*(2).

Vennix, J.A.M. (1996). *Group model building: Facilitating team learning using system dynamics*. Chichester, UK: John Wiley & Sons.

Walton, M. (1989). *The Deming management method.* London: Mercury.

Walton, M. (1991). *Deming management at work*. London: Mercury.

Ward P.T., & Mellor, S.J. (1985). *Structured development for real-time systems*. New York: Yourdon Press.

Waring, S.P. (1991). *Taylorism transformed.* Chapel Hill, NC: The University of North Carolina Press.

Wastell, D.G., White, P., & Kawalek, P. (1994). A methodology for business process redesign: Experiences and issues. *Journal of Strategic Information Systems*, *3*(1).

Weick, K.E., & Bougon, M.G. (1986). Organizations as cognitive maps: Charting ways to success and failure. In H.P. Sims, Jr., & D.A. Gioia (Eds.), *The thinking organization*. San Francisco: Jossey-Bass.

Wensley, A. (1996). Book review: Reengineering management. *Business Change and Re-engineering*, *2*(3).

White, T.E., & Fischer, L. (Eds.) (1994). *The workflow paradigm*. Alameda, CA: Future Strategies.

Wills, C. (1989). *The wisdom of the genes: New pathways in evolution.* New York: Basic Books.

Wills, C. (1993). *The runaway brain: The evolution of human uniqueness.* New York: Basic Books.

Woofford, J.C. (1994). Getting inside the leader's head: A cognitive processes approach to leadership. *SAM Advanced Management Journal*, *59*(3).

Yellen, R.E., Winniford, M., & Sandord, C.C. (1995). Extraversion and intraversion in electronically-supported meetings. *Information & Management*, *28*(1).

Yourdon, E., & Constantine, L.L. (1978). *Structured design: Fundamentals of a discipline of computer program and systems design.* New York: Yourdon Press.

Zuboff, S. (1988). *In the age of the smart machine: The future of work and power*. New York: Basic Books.

Zuboff, S. (1996). The abstraction of industrial work. In P.S. Myers (Ed.), *Knowledge management and organizational design.* Boston: Butterworth-Heinemann.

Glossary

ARPANET project. Large project initiated under the auspices of the Advanced Research Projects Agency (ARPA), a branch of the US Department of Defense (DoD), whose main goal was to build a network of shared mainframe-based computational resources by interconnecting major universities and research centers in the US.

Business process. A set of interrelated activities aimed at accomplishing an organizational task.

Business process improvement. The analysis, redesign, and subsequent change of organizational processes to achieve performance and competitiveness gains.

Business process integration. The electronic integration of several business processes involved in a company's supply chain; often going from ordering, passing through production, invoicing, inventory control, and ending with distribution and/or delivery.

Business process outsourcing. The farming out of entire business processes to external suppliers, building heavily on the infrastructure provided by the Internet.

Business process reengineering. Management movement pioneered by Michael Hammer and others, focusing on approaches for the radical improvement of productivity and quality of business processes, as well as the implementation of business process-centered organizational strategies; for example, business process outsourcing and organization-wide business process integration through the use of enterprise systems.

Compensatory adaptation hypothesis. Theoretical assumption that individuals who choose to use e-collaboration tools to accomplish collaborative tasks tend to compensate for the cognitive obstacles they perceive as associated with the lack of naturalness of those tools. This leads them to generate group outcomes of the same or better quality than those generated through the face-to-face medium.

Data. Carriers of information and knowledge that are either transferred or stored through a process of changing, or generating perturbations on, a given communication or storage medium.

E-collaboration. Electronic collaboration, or the carrying out of collaborative tasks employing electronic communication technologies. This is an umbrella term that comprises several other closely related fields, commonly known as computer-mediated communication, computer-supported cooperative work, groupware, group support systems, collaboration technologies, and, more recently, the so-called field of knowledge management.

E-collaboration paradox. Phenomenon characterized by two general and competing findings in connection with the impact of e-collaboration tools on groups. The first finding is that group members generally perceive face-to-face communication as posing fewer obstacles to effective communication than other communication media, particularly media generated by e-collaboration systems. The second finding is that when groups conduct collaborative tasks using e-collaboration systems to support interaction among group members, those groups often present the same or higher levels of performance as groups where members interact primarily face-to-face. This second finding is clearly contradictory with the first finding.

Establishment of computer networks. Period going from the late 1960s to the mid 1980s that began with a major development in 1967 (official start date according to most accounts), the ARPANET project, which provided the basis on which the now ubiquitous Internet has evolved.

Excellence movement. Management movement pioneered by Tom Peters and Robert Waterman comprising several best practices employed by successful companies in a variety of industries.

Expansion of local area networks. Period going from the mid 1980s to the early 1990s, which owes much of its existence to the development and widespread use of personal computers and their interconnection in local area networks.

First Industrial Revolution. Generally seen as the period from around 1770 to 1850.

Functionalism. Management doctrine pioneered by Henry Fayol comprising a set of prescriptions for structuring large organizations around forecasting, planning, and coordination activities.

Hawthorne effect. The business effect associated with the notion that, for the average worker, the desire to stand well with one's fellows and managers easily outweighs the influence of financial rewards and physical working conditions.

Information. Abstract entity that is carried by data, and that is eminently descriptive. From a linguistic perspective, the typical instance of information is the utterance called assertion. One example of assertion is: "Today is a sunny day."

Internet. Wide area network interconnecting many local area networks distributed around the world, which evolved from the initial infrastructure set in place through the ARPANET project.

Internet Era. Period that began in the early 1990s and that extends to the present day. This period is marked the development of the Web and the expansion of the commercial use of the Internet.

Internet stock bubble burst. Event that took place toward the end of the 1990s, when many companies whose business relied heavily on the Internet, or on other Internet-based companies, saw their market values skyrocket, only to see those values take an unprecedented nosedive in the early 2000s.

Knowledge. While information is eminently descriptive, and can refer to the past, present and future, knowledge is by its own nature eminently associative. That is, it allows us to associate different world states and respective mental representations, which are typically linked to or described by means of pieces of information.

Local area networks (LANs). Networks connecting several computers and computer peripherals usually located near each other (e.g., in the same building). The main appeal of LANs in the 1980s was that they enabled the sharing of what were then relatively expensive resources, such as laser printers and large-capacity/high-speed hard disks.

Mainframe. Large central computer connected to a number of "dumb" terminals – i.e., terminals with very limited or no processing capacity of their own (hence the term dumb).

Mainframe Era. Period marked by a dominance of large computer systems, usually known as *mainframes*, going from the early 1950s, with the emergence of the first mainframe assembly lines, to the late 1960s, with the first major computer networking projects.

Media naturalness hypothesis. Theoretical assumption that individuals who choose to use e-collaboration tools to accomplish collaborative tasks experience increased cognitive effort and communication ambiguity proportionally to the degree to which the tools suppress elements that are present in face-to-face communication (e.g., synchronicity, ability to convey/perceive non-verbal communication cues).

MetaProi. Group methodology for business process improvement; often referred to as a meta-process to indicate that it is a high-level process that describes how business process improvement should ideally be carried out in organizations. MetaProi is short for Meta-Process for Business Process Improvement.

Organizational development. The generic field of research and practice concerned with structural organizational changes that can have a positive impact on competitiveness.

Post-War Era. Period that goes from the end of World War II to the late 1980s.

Scientific management. Management school of thought pioneered by Frederick Winslow Taylor emphasizing business process improvement through the careful and precise measurement of the times and motions involved in relatively simple manufacturing activities.

Second Industrial Revolution. Period that goes from approximately 1850 to the years preceding the official start of World War II.

Task specialization. Notion proposed by Adam Smith, and also known as division of labor, stating that manifold gains in productivity could be achieved in manufacturing activities if every worker focused their efforts on one simple task of an assembly line.

Total quality management. Management school of thought pioneered by William E. Deming emphasizing continuous business process-focused improvement methods that build heavily on the use of statistics.

World Wide Web. An abstract collection of sites, created by Web servers (e.g., the server that runs Amazon.com's site), which uses the physical infrastructure provided by the Internet. Also known as WWW or the Web.

About the Author

Dr. Ned Kock, whose full name is Nereu Florencio Kock, Jr., is Associate Professor of MIS & Decision Science at Texas A&M International University (USA). He holds a BEE in electronics engineering, an MSc in computer science, and a Ph.D. in management with a concentration in information systems. Ned has more than 16 years of experience as a systems analyst and organizational development consultant. He is the author/co-author of several books, including the general interest book *Compensatory Adaptation: Understanding How Obstacles can Lead to Success*. He currently serves as Associate Editor for the *Journal of Systems and Information Technology* and Associate Editor for Information Systems for the journal *IEEE Transactions on Professional Communication*. Ned is a frequent speaker in international conferences and has received several awards and distinctions for his contributions to research and industry practice. Among those awards and distinctions are his inclusion in AcademicKeys' *Who's Who in Business Higher Education* (2003) and Manchester's *Who's Who of IT Professionals* (2002), and his receiving the CIGNA Research Fellowship (2000).

Index

A

action learning 16
action-oriented management researchers 11
Advanced Research Projects Agency (ARPA) 23
American Telephone and Telegraph (AT&T) 17
analog electrical waves 55
analysis 32
Argyris, C. 16
ARPANET 161
assembly line 12
automated procedures 73

B

bridges 27
bureaucratic procedures 73
business environment 89
business process 32
business process improvement 3, 32, 45, 132, 175, 213
business process improvement groups 175, 214
business process management era 18
business process outsourcing 19
business process redesign 121
business process reengineering 33, 132
business process-focused improvement 132
business processes 175

C

central data processing departments 2
central processing unit (CPU) 25
Champy, J. 34
chief executive officer (CEO) 2
client-server system 28
common predecessor relationship 40
common successor relationship 40
communication 8
competitive pressures 11
"computer folks" 2
computer networks 2, 23
computer processing 162
computer-mediated communication 2
computer-supported collaborative work 120
customer satisfaction 11

D

data 51
data buffers 44
data exchange 79
data flow view 41
databases 73
Davenport, T. 19
Deming, W. E. 3, 33
Department of Defense (DOD) 23
departmentalization 90
double-loop learning 16, 35

E

e-collaboration 1, 120, 132
e-collaboration paradox 7
e-collaboration system 214
e-collaboration technology support 6, 176
e-collaboration tools 6
e-mail 8, 53
economy 21
electronic digital computer 161
ENIAC 161
“excellence” movement 17

F

face-to-face communication 9, 120
face-to-face conversation 55
face-to-face interaction 8
face-to-face meetings 176
fax machine 25
Fayol, H. 13
file transfer protocol (FTP) 28
first industrial revolution 13
Ford 76
formation 213

G

General Electric 22
General Motors 14
group collaboration 1
group process 176
group support systems 120

H

Hammer, M. 3, 19
hardwired 120
Herzberg, F. 15
hypotheses 121

I

industrial era 13
industrial revolution 12
inflation 1
information 51
information era 56
information flows 89
information processing 73
information sharing 75
information society 56
information technology (IT) 51
international business machines (IBM) 23
Internet 21
Internet bubble burst 21
Internet era 26

K

Kanter, R. M. 18
knowledge 51, 73
knowledge communication 175
knowledge management 2
knowledge sharing 5, 74, 162
knowledge specialization 75, 90
knowledge transfer 77

L

local area networks (LAN) 24

M

MAF quality management 213
mainframe era 22
“management by walking around” (MBWA) 17
management consultants 11
management procedures 11
management thinking 1
managers 11

Maslow, A. 15
Mayo, E. 13
McGregor, D. 15
meta-process 176
MetaProi 175, 213
microcomputers 162
minicomputers 25
motivation 213
mutual fund management 90

N

National Science Foundation 28
neural networks 53
New Zealand 9
Nile River 12
Novell Corporation 28

O

organizational development 11
organizational functions 89
organizational knowledge 4, 74
organizational learning 16, 35
organizational processes 73
organizational productivity improvement 4

P

Pareto rule 32
Pentagon 21
personal computer 55
personal computers 22
Peters, T. 17
Pharaohs 132
Philadelphia 161
post-war era 15
predecessor-successor relationship 40
productivity 11, 176

Q

quality 11, 176

R

redesign 32
reengineering 20
Revans, R. 16
Roman Empire 12

S

Schon, D. 16
scientific management 1
second industrial revolution 14
securities analysis 90
single-loop learning 35
Smith, A. 12
specialization 4, 89
specialization of knowledge 89
structural organizational changes 11

T

Taylor, F. W. 13
total quality management (TQM) 1, 17, 32
total quality management movement 132

U

US Department of Defense 161
United States Army 161
University of Pennsylvania 161

V

volatile digital memories 55
Volkswagen 76

W

Waikato University 93, 213
Waterman, R. 17
Web browsers 28
Web servers 28
Web-based e-mail 29
Western Electric Company 14
Whitney, E. 12
workflow view 37
workflows 38
World Trade Center 21
World War II 14

www.ingramcontent.com/pod-product-compliance
Ingram Content Group UK Ltd.
Pitfield, Milton Keynes, MK11 3LW, UK
UKHW050946290926
15470UKWH00005B/1665

9 781591 403579